A GUIDE TO ENERGY MANAGEMENT

Inefficient energy use in buildings is both increasingly expensive and unsustainable. Indeed, the reduction of the energy consumption of existing buildings is as least as important as the design of new low-energy buildings. Controlling energy use is one thing, but it is important to assess or estimate it, and to understand the range of interventions for reducing its use and the methods for assessing the cost effectiveness of these measures.

This book should appeal to building managers and facilities managers and also to students on the increasing number of energy management modules in FE and HE courses. It clearly and concisely covers the various issues from a theoretical standpoint and provides practical, worked examples where appropriate, along with examples of how the calculations are carried out. It serves as a handbook for those carrying out energy management in owner-occupied buildings or working for FM organisations. It provides a template for instigating the energy management process within an organization. Guidance is also given on management issues such as employee motivation, and the book gives practical details on how to carry the process through.

Douglas J. Harris is a Lecturer in the School of the Built Environment at Heriot-Watt University, Edinburgh, UK.

A Guide to Energy Management in Buildings

Douglas J. Harris

Spon Press
an imprint of Taylor & Francis

LONDON AND NEW YORK

FIGURES AND TABLES

Figures

Tables

CHAPTER 1

BACKGROUND

The energy problem

The problems of global warming and climate change are widely acknowledged to be a product of man's profligate use of fossil fuels, and it is well known that there is an urgent need to cut our fossil fuel consumption substantially over the next few years. Approximately 50 per cent of our energy use, and carbon dioxide (CO_2) emissions into the atmosphere, are from the use of energy for heating, cooling and lighting buildings. Around 25–30 per cent is used in transport, while the remainder is used for industrial processing. At the Kyoto summit (December 1997), the United Kingdom made a voluntary commitment to reduce CO_2 emissions by 20 per cent by the year 2010, and other countries made similar commitments to reduce their use of fossil fuels. It is clear that reducing energy consumption in buildings can make a major contribution to achieving the reductions required, and good management of energy use in buildings is acknowledged as an important aspect of sustainable development. The importance of this is reflected in the Building Research Establishment Environmental Assessment Method (BREEAM) environmental assessment method for buildings, of which energy consumption is a major constituent. It has been estimated that businesses in Britain are losing £7 million a day through wasted energy in offices alone. Improved standards are required of new buildings – changes to Building Regulations Part L in 2010 meant that carbon emissions will have to be cut by 25 per cent compared with 2006 levels. Carbon reduction is no longer something that is just talked about, it has become an integral part of the building manager's job, and people are now scrutinizing carefully their energy consumption and carbon emissions, at home and at work. More ambitious recent targets require a reduction of 60 per cent in carbon emissions by 2050. This requires a huge change in our attitude to energy use and generation methods, and even with improvements in demand management we will need to generate 30–40 per cent of energy from renewable sources – amounting to an increase of about 1200 MW annually.

Reducing global warming by lessening our reliance on fossil fuels is therefore one reason for wanting to lower our consumption of energy in buildings. Countries such as China are expanding their economies rapidly, and with this expansion goes an increased demand for energy, much of it provided by fossil fuels such as oil, coal and gas. Another aspect of the energy problem is that of continuity of supply of the amounts of fossil fuel demanded; fuel reserves are dwindling, and many observers estimate that oil production is already past its peak – if demand continues to soar while supply falls, prices will inevitably escalate at a rapid rate, leading to severe economic and social problems. Thus, for economic and environmental reasons reducing energy use in buildings is imperative.

Recent global increases in oil and gas prices have raised public awareness about the cost of energy. In the United Kingdom gas price increases of 25 per cent in less than a year have led both domestic and business users to place a greater emphasis on energy efficiency. Globally, energy costs are following a similar trend, and although there may be periods of lower oil prices, the general trend will continue upwards. The reduction in our reliance on increasingly expensive fuel imports is a further driving force behind energy management.

The impact of current legislation on energy efficiency, the use of new materials and techniques for utilizing renewable energy, and the adoption of more energy-conscious design, mean that most new buildings will consume much less energy per square metre than the majority of those already existing. Initiatives such as the Code for Sustainable Homes and the Passivhaus standard help guide architects towards low-energy solutions, and it is the Government's aim to have all new housing zero carbon by 2016, not including imports from renewable

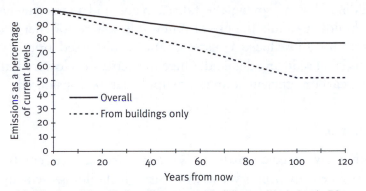

Figure 1.1 Effect on CO_2 emissions if every new building emitted at only 50 per cent of current levels without reducing energy use in existing buildings

energy sources. The effects of climate change in the long term will be to slightly reduce heating loads and increase cooling loads, but they are unlikely to have a significant effect on energy use. In most developed countries the rate of replacement of existing buildings with new ones is very low (about 1 per cent per year) and the time scale for substantial reductions in CO_2 emissions, in the absence of any work on existing buildings, is therefore very long (see Figure 1.1). The use of passive techniques such as greater thermal mass and better use of solar gains which help to keep consumption down in many new buildings is only applicable to a fraction of existing buildings, and in order to achieve a more rapid reduction in CO_2 levels, reducing energy consumption in existing buildings is consequently of major importance. Governments worldwide are instigating measures to improve sustainability, and energy efficiency is an essential element of all these initiatives.

Against this background, energy management attempts to analyse where energy is wasted in buildings and identify cost-effective solutions.

Energy use in buildings

The statistics paint a fairly gloomy picture, and highlight the urgency of carbon reduction measures. Energy management as described in this book is largely concerned with demand side management: that is, reducing the demand for energy in existing buildings by improving the fabric, plant and management of the building. While the use of renewable sources of energy can contribute to reduced emissions and is discussed briefly, the focus is on reducing demand. Once the demand is lowered, then more carbon-efficient sources of supply can be considered.

Domestic

Within developed countries in Western Europe, between 30 and 60 per cent of primary energy is used in buildings, chiefly to provide heating, cooling and lighting. In Britain, approximately half of this is used in housing. The distribution of energy use within different building types varies greatly. In housing, approximately 58 per cent of delivered energy is used for space heating, and space and water heating combined amount to 82 per cent of consumption. Since 1970 energy use for space heating has risen by 24 per cent, for water heating by 15 per cent, and for lighting and appliances by 15 per cent. Energy use for cooking has fallen by 16 per cent. Overall electricity consumption has risen by 63 per cent in the same period. Overall electricity consumption has risen by 63 per cent in the same period, partly due to an increase in the use of home computers and electronic games and other entertainments.

Some of these changes are driven by changes in lifestyle and demographics – the number of single-occupancy households has risen sharply in the last twenty years. Central heating, which was used in only one third of houses in 1970, is now installed in 82 per cent of dwellings. There has been a significant change in the type of housing built in the past twenty years: more one and two-bedroom flats, for example, to meet the changes in family structure. In spite of the large number of houses built, it remains the case that 40 per cent of the UK housing stock was built before 1945, and 46 per cent between 1945 and 1984, leaving roughly 14 per cent built since 1984. Much of this housing is unimproved in terms of energy efficiency.

The greater use of central heating has resulted in higher internal temperatures – average 18°C in 2000 as opposed to 13°C in 1970, and homeowners now tend to heat more rooms in their houses. The effect of this on national energy consumption is significant, particularly when one remembers that a 1°C rise in room temperature equates to about 7 per cent increase in energy consumption. While approximately 70 per cent of houses have loft insulation, only about 19 per cent have their cavity walls insulated, 69 per cent of dwellings having cavity walls. Around 39 per cent of all dwellings have some or all of their windows double glazed.

The average home energy use per year is approximately 20,500 kWh of gas and 3,500 kWh of electricity, costing £531 and £691 respectively. Consumption in kWh is split as follows:

Space heating	58%
Water heating	24%
Total Electricity	18%
Lighting	3.4%

It has been estimated that it would cost the average homeowner £2000 to reduce their energy consumption by 25 per cent. On the basis of the above figures, this would produce an attractive payback period of just over two and a half years. Most of the techniques proposed rely on reducing the load and using more efficient appliances from conventional mains supplies, but the use of renewables is increasing. By April 2011 there was approximately 77 MW of installed photovoltaic (PV) power in the United Kingdom, and the number of PV installations has risen rapidly since the introduction of feed-in tariffs. Long payback periods still deter householders from investing in these technologies, though, and small PV and wind turbine systems still have paybacks in excess of 25 years.

Non-domestic

Non-domestic buildings cover a wide range of building types and uses. The most common are offices, and their energy consumption has been the best documented, but other building types produce substantial emissions. Hospitals are the largest emissions producers, with an average of 4089 tonnes of CO_2 per year, while prisons are second with 2849 tonnes. Typical figures for offices are shown in Table 1.1. As sizes vary greatly, typical consumption figures are given in kWh/m^2 of treated or usable floor area.

Table 1.1 Typical office energy consumption and emissions in the United Kingdom

Naturally ventilated cellular	Energy kWh/m²	Kg CO2/m²
Heating and hot water	160	9
Cooling	0	0
Lighting	25	4
Office equipment	30	6
Total	215	19
Air conditioned		
Heating and hot water	180	8
Cooling	30	4.5
Lighting	50	8
Office equipment	30	6
Total	290	26.5

After lighting, computers and monitors have the highest energy consumption in the office. Studies show that about half are left on overnight and at weekends – about 75 per cent of the hours in the week. Simply turning them off at night can result in substantial savings. As an incentive to reduce emissions, it seems likely that the Display Energy Certificate (DEC) scheme, which at present covers only public-sector buildings, will be extended to all non-residential properties.

Aims and objectives of energy management

Energy management in buildings is concerned with maximizing the use of energy resources while providing the desired environmental conditions and services inside the building at the least cost. Energy management can also be applied to industrial processes, particularly those requiring the use of heat or steam, but here we are mainly concerned with the use of energy in buildings to provide thermal, visual and acoustical comfort for the occupants. The range of building types that an energy professional may be required to examine include homes, offices, retail premises, sports centres, cinemas, theatres, restaurants, schools, universities, and other commercial premises.

The key aims of energy management are:

- minimizing energy consumption
- optimizing size of plant
- maximizing energy efficiency
- minimizing energy waste
- reducing carbon emissions
- reducing costs.

Energy management involves the measurement or estimation of the energy consumption of a building, the provision of recommendations for improvement, and the development of a strategy for maintaining continuous improvements in the energy performance. Typical techniques for reducing energy consumption will be discussed in detail later. They include:

- use of thermal insulation
- recovery of waste heat
- combined heat and power (CHP)
- the use of more effective heating and cooling systems
- the use of natural methods of ventilation and cooling
- better use of controls for heating, cooling and lighting
- a strategic approach to managing and reporting energy use.

Energy costs

The running costs of many organizations tend to be dominated by the cost of labour, with energy accounting for only 5–10 per cent. When savings are needed it is much easier simply to reduce staff numbers – which also reduces a company's capability. If a company's energy bill is considered as a total rather than as a percentage of running costs, the amounts involved can be considerable, and for energy-intensive industries with a small workforce, such as the Scotch whisky industry, it may represent up to 50 per cent of the running costs. Furthermore, sudden sharp increases in fuel prices can cause serious cash flow problems for small and medium-sized businesses. Energy efficiency struggles under the handicap that energy costs are often 'invisible', and it does not create an income. When energy costs are reduced there is no visible increase in the income stream, often only a record of what would have been spent if energy-saving measures had not been introduced.

Example

Consider a 25-year-old air-conditioned office building situated in the United Kingdom, 100 m by 20 m with three stories, with a total annual energy consumption of 300 kWh/m² (i.e. a fairly inefficient, but not atypical, building).

Total floor area = 6000 m²
Total annual energy consumption = 100 × 20 × 3 × 300 =
    1,800,000 kWh per year
Cost at average fuel price 8p/kWh = £144,000 per year
Or £24 per sq. metre per year.

With good energy management, 10–30 per cent per year could be saved, i.e. £14000–£42,000. Individual cases may show larger savings for a relatively small outlay.

If we consider the total amount spent on energy by an organization in a year, then we can see that energy management can be worthwhile, and many savings can be made at little or no cost.

Small inexpensive changes may have a substantial effect if considered nationwide – a 2°C lowering of temperature in winter in all buildings in the United Kingdom would save 11 x 10^{16}J of energy, about £900 million.

Need to comply with legislation

One of the means by which governments endeavour to lower their national energy consumption and carbon emissions is legislation. In the United Kingdom this includes the Building Regulations, and in other countries there are likely to be various forms of building code which must be complied with. These tend to influence building design, as they specify such variables as U-values, maximum energy consumption per square metre, the efficiency of building services equipment such as boilers, air conditioning systems and lighting, and controls. While many of these regulations apply only to new build, gradually increasing constraints are placed on refurbishments, building extensions and adaptations. These regulations will only become stricter in the coming years.

Cost-effectiveness of energy-saving measures

Part of the job of an energy manager is to optimize the available resources. When intervention to save energy is considered, many factors must be taken into account: not only the energy savings, but the cost-effectiveness of the energy-saving measures. We need to consider how much energy might be saved, the capital investment required, the payback period, and for how long the measure will be effective. For example, it is easy to reduce the heat losses from a building merely by adding more and more insulation to the walls. Although the addition of insulation saves energy, its cost may be significant, and we have to consider how much energy it really saves over the lifetime of the building, and whether a better investment might be made in more efficient plant or renewable energy sources. A simple illustration of this is given in Chapter 2, while techniques for assessing the cost-effectiveness of measures are examined more closely in Chapter 3.

CHAPTER 2
ASPECTS OF BUILDING ENERGY USE

Environmental requirements in buildings

The energy requirements of a building are largely dependent on the needs of the building occupants and the activities taking place there. The provision of comfort for the occupants is one of the primary functions of a building. Aspects of comfort include:

- **Thermal** – a state of thermal equilibrium whereby the occupant feels neither too hot nor too cold. The factors involved are air temperature, radiant temperature, relative humidity and air movement. In well-insulated buildings the radiant and air temperatures tend to be very close, and people feel fresher if the air temperature is slightly lower than the radiant temperature. Generally people in offices and similar buildings are happy with operative temperatures 19–25°C, relative humidity 40–70 per cent, and a small degree of air movement, generally below 0.15 m/s. Other factors affecting the comfort temperature include uniformity of temperature distribution, level of activity and clothing. For these reasons, different conditions are required in gymnasia, sports centres and swimming pools. Both domestic and office temperatures have risen gradually over the years as people's style of dress at home and at work has changed. Average temperatures during working hours in offices are commonly 22°C in winter, and sometimes more; air-conditioned spaces often feel too cool in summer. Allowing a greater range of temperatures in summer and winter would lead to large savings in the long term, at no cost.

- **Visual** – occupants should be able to carry out their necessary tasks without visual strain. The quality of lighting is as important as the quantity – light should be of appropriate colour, glare should be eliminated, and lighting should be suitable for viewing computer screens. In addition, lighting for safety (e.g. exit in event of fire or power cut) should be provided. Higher lighting levels than needed waste energy and can lead to eye strain and headaches, if endured over long periods.
- **Acoustic** – provision of appropriate acoustic environment, without intrusion of excessive noise from neighbouring rooms or buildings, or from the building services equipment, e.g. fans, pumps, lifts.

The requirements for different building types can be found in the appropriate national standards and regulations, or in the Chartered Institute of Building Services Engineers (CIBSE) and American Society of Heating, Refrigerating and Air-conditioning Engineers (ASHRAE) guides.

Most of the information in this guide concerns technical solutions to the problem of saving energy, but even before those solutions are applied there is much that can be done, such as changing people's attitudes and expectations. For example, insulating a house may not lead to energy savings, because the owner may choose to use the same amount of energy as before but enjoy a much higher temperature; similarly, two identical adjacent houses may have energy consumption levels that differ by 100 per cent, because of the lifestyle and behaviour of the occupants.

There is an ongoing debate concerning thermal comfort, and it is a widely held view that the comfort levels described above are too narrow in range, and do not take account of the ability of people to adapt to their surroundings. Widening the limits considered acceptable, and the use of appropriate levels of clothing, would permit a substantial amount of energy to be saved on heating and cooling, without incurring any costs.

Where and how energy is used in buildings

Energy is used in a number of ways in buildings, and the most cost-effective interventions result from targeting the sectors where most energy is used. This clearly depends on the building type, construction and building services details, and the climate in which the building is located.

In the United Kingdom most of the energy used in houses is used for space and water heating, while in offices a much greater proportion is used in lighting. In hot-dry climates such as prevail in much of the Middle East, up to 90 per cent of the energy consumption may be accounted for by air conditioning,

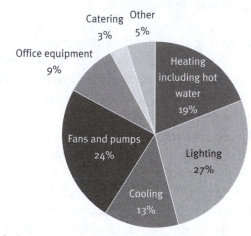

Figure 2.1 Typical energy consumption pattern for a 20-year-old air-conditioned office building in the British climate

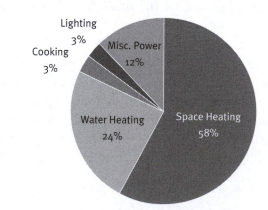

Figure 2.2 Typical domestic energy consumption pattern for a private house in the British climate

even in housing, whereas air conditioning is virtually unknown in British houses.

The heating requirement is divided into two elements, space heating and water heating. The space heating is further divided into two components, fabric losses and ventilation losses. All buildings require ventilation, and whether this is supplied mechanically or naturally, it incurs a heat loss on the building.

The heat gains in summer can be divided into a number of elements: heat gains through the building fabric, heat gains through ventilation air, heat gains from internal sources such as people, and those from electrical equipment such as lighting, computers and other office equipment.

Energy balance

A simple heat balance can be used to understand the heat flows though a building. Energy may enter and leave the building, and may also be stored in the fabric and contents.

Energy in = Energy out + Energy stored

While there are time lags due to absorption of heat in the building materials to varying degrees, when considered over a long enough period the residual energy storage is small compared with the flows in and out, and a 'steady-state' energy flow situation may be assumed.

ENERGY IN = ENERGY OUT

The time lags will affect heat-up and cool-down periods in winter and summer, and comment is made on this in Chapter 8 when discussing optimum start and stop, but on a first analysis, it can be assumed that all the heat that enters the building will leave through the fabric, through ventilation air, or through the cooling system, if installed.

Energy consumption and heat gains of electrical equipment

Much of the energy used in a building is in the form of electricity, which powers lighting, computing, photocopiers, entertainment, electric motor refrigerators, washing machines and so on, and is ultimately converted into heat. While in winter these heat gains have a beneficial effect in that they offset the amount demanded of the heating system, in summer they are undesirable as they increase the cooling load of the building.

It is unlikely that sufficient metering will be installed to identify and quantify all the end-use applications of electricity; monitoring exercises may reveal some information, but in the case of most buildings at least some of the electrical gains will have to be estimated.

For equipment with a constant consumption it is a matter of estimating the hours used and multiplying by the power rating.

$$E_e = W_r \times h$$

Where

E_e is the power consumed
W_r is the power rating
h is the number of hours used.

Items such as computers consume a variable amount of power depending on their state; a laptop may use 25 W in sleep mode, 100 W working on screen-save, and 200 W with the screen on and hard drive in use.

In order to minimize energy consumption from appliances, the most effective approach is to specify low-energy appliances in the first instance and use them for the minimum time. Low-energy appliances can be identified by the 'Energy Star' system (US) and Energy Ratings system used in the United Kingdom and the European Union. In the latter scheme a label specifies the consumption and indicates a rating of A–G (A** is now available for refrigerators) in descending order of efficiency.

Internal heat gains also include those from the occupants, varying from just over 100 W for sedentary activities to over 400 W for heavy manual work.

The installed wattage of lighting systems can usually be calculated by multiplying the number of lamps or tubes by the individual rating, and making an estimate of the hours used.

Example: a large open-plan office has 180 fluorescent tubes of 54 W each. Calculate the annual energy consumption.

Installed wattage = 54 × 180 = 9.72kW.
Hours of operation: 5 days a week, 50 weeks a year.
8.00 am–18.00 pm.
It is estimated that the lighting is on 80 per cent of the office hours.

Total hours per year = 10 × 5 × 50 × 0.8 = 2000 hours.
Total consumption 9.72 × 2000 kWh = 19440 kWh.

Reductions in energy consumption can be brought about by using more efficient light sources to produce a lower installed wattage and reducing the number of hours of operation at full output.

If counting the light fittings and their individual ratings is not feasible, then an estimate should be made. Heat gains from lighting vary between approximately 8 W/m² for efficient systems and 20 W/m² for poor ones, the overall consumption being a function of working hours and the type of controls used.

If the rating of computers and other office equipment is known, then adding the individual gains is a simple matter; in the absence of this information, allowances for other office equipment range from 5–15 W/m².

It is unlikely that all the equipment and all the lighting will be in constant use in a building. In order to allow for intermittent use, diversity factors are used. A 200 W printer in use for 8 hours would produce a total heat gain of 1600 Wh. Applying a diversity factor of 0.6 reduces the heat gain to 960 W.

Heat gains from cooking appliances are more difficult to assess, particularly in large non-domestic kitchens, because exhaust hoods remove a significant proportion of the heat directly. These are latent and convective gains; most of the radiant gains enter the room.

Example of heat loss calculation

The U-value is a measure of the heat transfer through a building element such as a wall and is measured in W/m^2K. It can be considered to comprise three main thermal resistances: the inner surface resistance, the resistance to conduction of the wall itself, and the outer surface resistance. The surface resistances can be obtained from the CIBSE and ASHRAE guides; the wall resistance is the sum of the resistances of the individual layers; the resistance of each layer is simply the thickness t_n divided by its thermal conductivity k_n.

$$U = \frac{1}{R_t}$$

$$R_t = R_{si} + R_w + R_{so}$$

$$R_w = R_1 + R_2 + \cdots$$

$$R_n = \frac{t_n}{k_n}$$

For a single-leaf brick wall in a 'normal' location (neither highly sheltered nor highly exposed) the following values are used:

Without insulation

t_w = 105 mm, k_w = 0.44
R_w = 0.105/0.44 = 0.238
R_{si} = 0.123 m^2K/W
R_{so} = 0.055 m^2K/W
R_t = 0.123 + 0.238 + 0.055
= 0.417 m^2K/W
U = 2.4 $Wm^{-2}K^{-1}$.

With insulation

Adding 50 mm of insulation with thermal conductivity k = 0.035 gives:

R_{ins} = 0.05/0.035 = 1.428
R_t = 0.123 + 0.238 + 1.428 + 0.055 = 1.845
U = 0.54 $Wm^{-2}K^{-1}$.

Adding a further 50 mm of insulation gives;

R_{ins} = 0.1/0.035 = 2.856
R_t = 0.123 + 0.238 + 2.856 + 0.055 = 3.273
U = 0.305 $Wm^{-2}K^{-1}$.

It is clear from the example that the benefit of adding insulation is subject to a law of diminishing returns. The first 50 mm of insulation reduces the U-value by 1.86, while the second 50 mm reduces it by only 0.23. Figure 2.2 shows the effect on the U-value of adding different thicknesses of insulation to a plain brick wall. Considering the energy savings (in terms of energy cost) over the lifetime of the insulation, and balancing that against that the cost of purchasing and fitting the insulation, it becomes evident that beyond a certain point we are investing large amounts of capital for little effect on running costs. Complex formulae have been developed to estimate the economic thickness of insulation in different circumstances, but one hindrance to using them is that although present energy and insulation costs may be known, it is difficult to predict how the energy cost will change. A sizeable change in fuel costs has a profound effect on the economics of insulation; as the fuel price increases relative to installation cost, the optimum economic thickness increases. Even for low insulation cost and high energy cost, adding more than 150–250 mm of insulation to an existing element is rarely a worthwhile investment.

Example of insulation payback calculation

Consider a detached bungalow located in Eastern Scotland, with floor plan 10 m by 7 m, walls 3 m high, with glazing constituting 40 per cent of the wall area, of single-leaf brick construction, with no insulation. The number of degree days at this location is 2234.

The U-value of the walls is 2.4 $Wm^{-2}K^{-1}$.

Using the simplified version of the degree-day method (described in the Appendix) the Annual Energy Consumption (AEC) to compensate for heat losses through the walls only, with a 60 per cent efficient boiler, is given by:

$$AEC = 2.4 \times 61.2 \times 2234 \times 0.024 \times 0.7/0.6 = 9187 \text{ kWh}$$

Which at 5 p/kWh gives an Annual Energy Cost (AEc) related to the walls of

$$AEc = £460/year.$$

Adding 50 mm insulation reduces the U-value to 0.54 $Wm^{-2}K^{-1}$.

$$AEC = 0.54 \times 61.2 \times 2234 \times 0.024 \times 0.7/0.6 = 2067 \text{ kWh}$$

giving an AEc of £103/year.

This results in savings of £357/year. Adding 50 mm insulation costs £8/m² (£490) and has a payback period of 1.37 years.

Alternatively, if 100 mm insulation had been added from the outset the energy consumption would be

$$AEC = 0.305 \times 61.2 \times 2234 \times 0.024 \times 0.7/0.6 = 1167 \text{ kWh}$$

giving

$$AEc = £58/\text{year}$$

savings of £402 over the original wall. The cost of the insulation and installation is approximately double (£980), giving a longer payback period of 2.4 years.

In this instance the thicker insulation gives a longer payback period and may therefore appear less attractive; however, payback period is only one of the criteria that may be used to make a decision.

The effect of insulation on energy costs and payback depends on the nature of the initial wall construction. For a brick cavity wall without insulation, the figures become:

Original, no insulation, $U = 1.38$ Wm^{-2}K^{-1}, energy cost £264
Total 50 mm insulation, $U = 0.31$ Wm^{-2}K^{-1}, energy cost £59,
    payback period 2.4 years
Total 100 mm insulation, $U = 0.22$ Wm^{-2}K^{-1}, energy cost £42,
    payback period 4.4 years.

The better insulated the wall already is, the longer the payback period for the insulation.

In the examples given the payback periods are relatively short, and the measure would be expected to last for 20 years. Therefore the total net savings (i.e. savings minus cost of insulation) in each case over 20 years are as follows.

Single brick, 50 mm £6650
Single brick, 100 mm £7060
Cavity brick, 50 mm £3610
Cavity brick, 100 mm £3460

In the former case the net savings increase with increasing insulation thickness, whereas in the latter, which is already better insulated, they decrease.

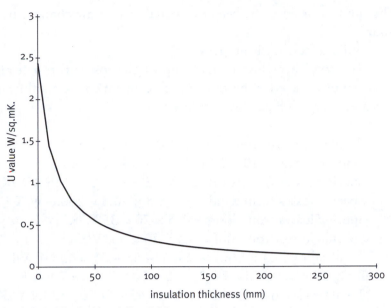

Figure 2.3 Effect of insulation thickness on U-value

Once the specific heat loss coefficient and annual heating energy consumption have been determined, steps to identify the appropriate energy-conserving measures can be taken. Firstly, the routes of greatest heat loss should be identified. If information on ventilation rates and the U-values of the elements of the outer envelope of the building is available, the heat loss may be broken down into its individual components – boiler inefficiency, ventilation or infiltration, and loss through the building fabric. The latter can be further broken down into walls, windows, roof and ground floor.

Later chapters will show how some of these measurements or estimates can be made, but for the present we will assume the data below.

Continuing with the example of the bungalow used earlier, the U-values of the outer envelope of the building are as follows;

Walls	2.4 W/m²K
Windows	5.8 W/m²K
Ground floor	0.5 W/m²K
Roof	0.5 W/m²K

The infiltration rate has been estimated at two air changes per hour.

Boiler efficiency is 60 per cent.

The total heat loss is made up of the losses through all these routes. The contribution of each element to the total heat loss and energy bill can be calculated.

Wall area = $(10+10+7+7) \times 3 \times 0.6 = 61.2$ m^2
Window area = $(10 + 10 + 7 + 7) \times 3 \times 0.4 = 40.8$ m^2
Specific losses from walls, $\Sigma UA = 2.4 \times 61.2 = 146.8$ W/K
Specific losses from windows = $5.8 \times 40.8 = 236.6$ W/K
Specific losses from floors = $0.5 \times 70 = 35$ W/K
Specific losses from roof = $0.5 \times 70 = 35$ W/K
Total fabric loss = $146.8 + 236.6 + 35 + 35 = 453.4$ W/K
Infiltration = $0.33nV = 0.33 \times 2 \times 10 \times 7 \times 3 = 70$ W/K
Total fabric plus infiltration losses = $453.4 + 70 = 523.4$ W/K

Now the proportion of the total loss through each element can be calculated.

Walls = $146.88/523.52 = 28\%$
Windows = $236.64/523.52 = 45\%$
Floors = $35/523.52 = 6.75\%$
Roof = $35/523.52 = 6.75\%$
Infiltration = $70/523.52 = 13.5\%$

These figures enable the areas for improvement to be pinpointed.

Putting the above in order of descending energy cost:

Windows	45%
Walls	28%
Infiltration	13.5%
Roof	6.75%
Floor	6.75%

The elements that need to be targeted for improvement are those showing the greatest heat loss, the windows and walls. While the U-value of the roof is not particularly good and may be relatively simple to improve, as only 6.75 per cent of the heat loss occurs through that route it is unlikely to be cost effective. For relatively simple buildings a spreadsheet model can be produced to carry out calculations of this kind.

The heat loss through a wall can be reduced progressively by adding more and more insulation, but there is a limit to the useful thickness of insulation that may be applied, which depends on a number of factors, including cost, fuel price, climate, the building construction and its use.

Figure 2.4 shows a number of scenarios: low energy cost, high-energy cost, low insulation cost and high insulation cost. Even for low insulation cost and high energy cost, beyond 150–250 mm of insulation, the investment is not worthwhile in financial terms, while for low energy cost and high insulation cost, the optimum insulation thickness is about 30 mm in this example. The difficulty for the energy manager is that although present energy and insulation costs may be known it is difficult to predict how the energy cost will change, although we can be fairly certain that it will continue to increase. A sizeable change in fuel costs has a profound effect on the economics of insulation, as Figures 2.4 and 2.5 show.

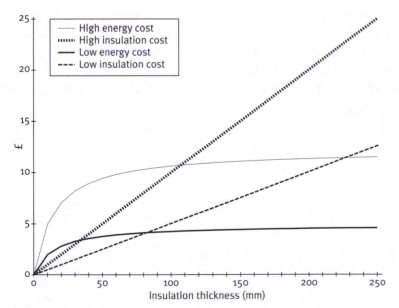

Figure 2.4 Sample scenarios

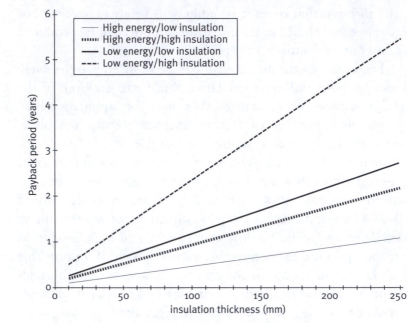

Figure 2.5 Payback period in relation to insulation thickness

Although some of the energy-consuming (and heat-producing) activities in the building may be assessed from power ratings, it may be necessary in some cases to make measurements of conditions and energy use, and these are explored later in Chapter 5. Investigation into the detailed energy exchanges in the building is pursued during the energy audit, which is described in the next chapter.

CHAPTER 3
ENERGY AUDITS

The energy audit

An energy audit is an essential tool of energy management. It is an investigation and detailed analysis of the energy entering and leaving a building, and is carried out in order to pinpoint those areas where improvements in the building fabric, services, controls and management can be made. The main aim of the energy audit is to identify actions that will lead to savings in energy and costs. Other aims include;

- reduction in carbon emissions
- improved environmental conditions for the occupants of the building
- the development of a system for recording energy use
- the development of monitoring and targeting schemes.

The energy entering the building comes from a number of sources; fuels, such as oil, coal, gas and electricity, and 'free' sources, such as solar radiation and people, which supply light and heat. All this energy is eventually converted to heat and leaves the building via a number of routes:

- transmission through the fabric
- infiltration and ventilation losses
- flue losses
- cooling tower losses.

The energy is used to power:

- heating equipment
- lighting
- cooling equipment
- fans
- pumps
- cooking appliances
- refrigerators
- small power (computers, copiers, fax machines, home entertainment etc)
- larger items of machinery
- lifts and escalators.

The energy audit process may be divided into a number of phases:

- Pre-survey information and data collection
- The building survey
- Analysis of the data collected
- Formulation of energy-saving solutions
- Reporting of results.

Pre-survey data collection

It is important to collect as much information as possible about the building before embarking on a walk-round survey. A number of sources of information may be available. The utility bills for the building will yield useful information on the amount of energy purchased and the tariffs paid; where a number of buildings are supplied and paid for on the same bill, readings from any sub-meters should be used in addition. If sub-metering is not available, then this may be noted as a potential area for improvement. Sub-meters may not be linked to billing, but identifying the consumption related to individual buildings is important. Even if a meter is not used for billing, regular readings, taken manually if necessary, should be incorporated as part of the monitoring strategy. Electricity bills will also show maximum demand charges (see Chapter 8), and further analysis should reveal whether load-shifting is feasible as a means of reducing these charges. A series of bills going back a number of years will also enable long-term trends in energy usage to be revealed.

Any plans, elevations or technical data on the building should be obtained. These will give useful information on dimensions, construction materials, (possibly U-values) and important information on the building services plant, including type, size and control strategy.

The information collected during this phase may include:

- the utility bills: gas, electricity, oil, solid fuel, water
- plans and elevations of the building, including building services systems
- location of the building, climate
- information on the controls and building management system (BMS)
- information on the structure of the building, U-values and materials
- information on the building's purpose, hours of work and operation.

Walk-through survey

The pre-survey data collection will give a general idea of the size and layout of the building and the type of plant used. However, a walk-through survey of the building is essential, and will provide much additional information which cannot be gleaned from the plans. One purpose of the walk-through is to determine whether the building is as stated on the plans; often changes are made such as extensions, new ventilation plant, additional insulation, and the like, which are not noted on the plans available. The condition of the building should be noted too: fabric defects (including draughts caused by badly-fitting windows and the like) contribute to energy waste, as does missing or damaged lagging on pipes or insulation on roofs and walls. The standard of cleanliness of light fittings is also to be noted – if lighting output is related to daylight level, dirty fittings may result in 20 per cent more energy being required.

As far as possible you should determine whether the U-values are as stated. This may be difficult or indeed impossible to ascertain without damaging the building fabric, but interviews with building managers, engineers, caretakers and others can provide useful anecdotal evidence to help build up a complete picture of the nature and operation of the building.

Detailed inspection of plant rooms is essential to confirm the type of plant, ratings, potential efficiency, and to provide a visual indication of the standard of maintenance. Any specific problems such as broken valves, leaks, missing lagging and other defects should also be noted here. In an older building, important items of plant such as boilers may have been replaced and the information given on the plans will be out of date. The location of meters and sub-meters should be identified.

Conditions in the building may be measured during the walk-through. A large range of hand-held devices is available to enable spot reading of temperature, humidity, lighting level and CO_2 level to be taken. These will only give an indication of the conditions prevailing at the time of the survey, but can be useful in identifying potential problems, for example with temperature control.

Other control issues may also be identified during the visit:

- Do the users have manual override for heating controls?
- Are there open windows immediately above working radiators?
- Can the users adjust light levels?
- Where are the light switches?
- Are computers switched off at night?
- Is there manual control of ventilation?
- Are fans switched on at night for night cooling?

There may be opportunities (with the employer's permission) to interview the building users or issue questionnaires. In the absence of extensive sub-metering, information to help with the assessment of individual items of energy consumption should be collected, noting what items of plant are running, whether lights are on, and what the policy is on such matters.

Analysis of the data collected

The form and extent of the analysis carried out depends on a number of factors: the depth of the audit, the nature of the building, the degree to which the energy consumption data can be disaggregated into end-uses, and the needs of the client.

It may be that sufficient information is obtained from the pre-survey and the walk-through to make a detailed energy analysis of the building. Alternatively, additional data may be acquired through a monitoring exercise. The equipment required is described in Chapter 4.

For large, complex buildings it may be worthwhile to model the performance using proprietary software. A number of packages are available of varying complexity, allowing simple steady-state heat loss calculations or detailed hourly simulations of HVAC performance to be made. If reasonably accurate input data is available, it is possible to simulate the effect on energy consumption of changes to the building such as upgrading the fabric insulation and improving plant efficiencies.

Where possible the measured or estimated energy consumption should be compared with that of other buildings having the same function. The Carbon Trust issues a large number of publications outlining typical and best-practice energy use patterns in a large range of building types, including offices, academic buildings, theatres, health care buildings and many others. Comparisons are normally based on $kWh/year/m^2$.

Formulation of energy-saving solutions

The analysis of energy consumption in the building will identify where energy use is high and where there is waste. The means by which waste is reduced are various, and range from no-cost solutions such as changes to occupant behaviour, through low-tech solutions such as adding blinds to the windows, to highly engineered and costly measures such as the installation of a combined heat and power plant. Examples of such measures are given in Chapter 4. When suitable means have been identified, the estimated savings should be calculated, and their cost effectiveness assessed.

Reporting

Finally, a report can be made to the client outlining;

- the present state of the building
- an analysis of current energy use

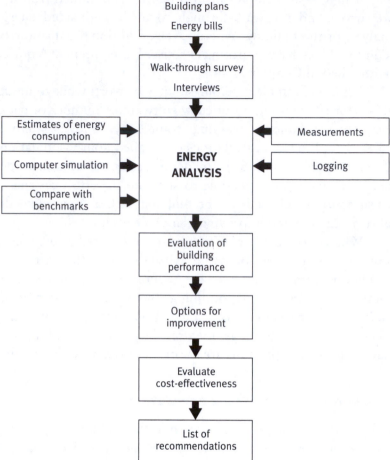

Figure 3.1 Summary of energy audit process

- identification of areas of waste and where energy can be saved
- details of the kinds of intervention which will reduce energy use
- details of the savings possible
- the cost-effectiveness of the methods recommended.

The measures may be listed in order of preference using the criteria specified by the client, such as lowest cost, shortest payback, highest net present value (NPV), as required. The energy audit process is summarized in Figure 3.1.

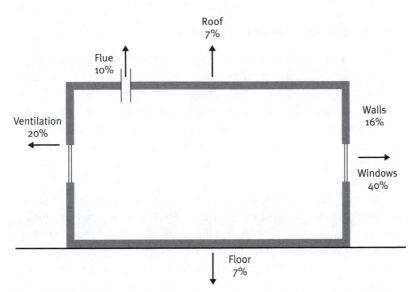

Figure 3.2 Distribution of annual heat losses from a typical house in the British climate

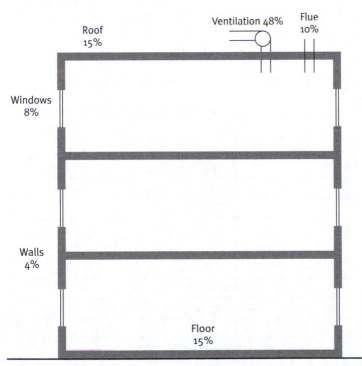

Figure 3.3 Typical heat losses from an office building in the British climate

Example of analysis of heating data

Different types of building use energy in different ways, and even heat losses from the fabric show marked variations (see Figures 3.2 and 3.3). If the heating of a building in winter is correctly controlled, there should be a strong correlation between energy use and external temperature, therefore a plot of energy use against degree days should yield a straight line. Such a plot, however, shows only that consumption increases in winter, but reveals little in the way of detailed analysis.

An example of how raw energy use data may be analysed using Degree days (see Appendix 1) is given below. It is based on a building in the UK climate which needs to be heated for about half the year. In this instance mains gas is used for both heating and cooking, but not for domestic hot water, and monthly bills are available. Also, monthly heating degree day data is available for this location. Utility bills tend to be issued on a quarterly or monthly basis, but if a BMS is installed or manual readings are taken, then weekly or even daily data may be available for a much finer level of detail. Large users of energy in the United Kingdom may require automated half-hourly metering in order to negotiate appropriate tariffs with the suppliers.

Table 3.1 Monthly gas consumption

Month	Total gas consumption (m³) over period
Jan	932
Feb	824
Mar	642
Apr	461
May	83.6
Jun	91.7
Jul	80.3
Aug	74.1
Sep	86.7
Oct	558
Nov	734
Dec	786
Annual total	5853.4

Table 3.2 Monthly energy consumption and degree days

Month	Total energy consumption (kWh) (gas)	Degree days
Jan	10,123	576
Feb	8,963	453
Mar	6,982	324
Apr	5,015	272
May	909	0
Jun	997	0
Jul	873	0
Aug	806	0
Sep	943	0
Oct	6,063	327
Nov	7,987	434
Dec	8,546	524
Annual total	49,661	2,910

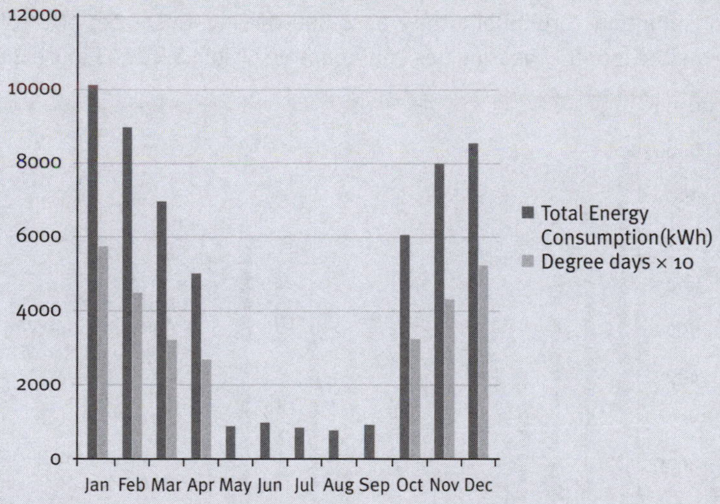

Figure 3.4 Energy consumption and degree days (× 10)

Table 3.1 shows the monthly gas consumption for the building. The gas consumption in cubic metres is divided by its calorific value (39.1MJ/ m³) to convert to kWh. Table 3.2 and Figure 3.4 show the kWh consumption along with the degree days at the site for the same period. If a BMS is used, external air temperatures may be recorded and accurate values of the local degree days can be calculated; alternatively, degree day values may be obtained from a number of sources. Gas is used for both space heating and cooking, but the amount used for each end-use in any given period is unknown. Water heating is carried out separately, by electricity. The values in Figure 3.4 include both heating and cooking energy; separation of the two enables a useful plot to be made. As only one meter is installed covering the total gas use, the data must be manipulated in order to separate it into the two end-uses. During the months May to September inclusive the heating is switched off entirely and no heating can be used in the building, even if the conditions are cool enough to require it. Therefore the gas consumption during these months is entirely due to cooking. As the amount of energy used for cooking does not change significantly with the season, taking the average gas consumption during the summer months will give a reasonable estimate of the monthly cooking energy consumption throughout the year. Subtracting this from the total gives the monthly heating gas consumption (Table 3.3 and Figure 3.5).

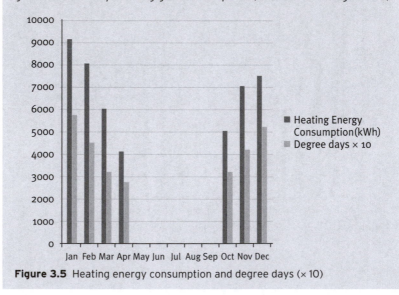

Figure 3.5 Heating energy consumption and degree days (× 10)

Table 3.3 Monthly heating energy consumption and degree days

Month	Heating energy consumption (kWh)	Degree days
Jan	9,217	576
Feb	8,057	453
Mar	6,076	324
Apr	4,109	272
May	0	0
Jun	0	0
Jul	0	0
Aug	0	0
Sep	0	0
Oct	5,157	327
Nov	7,081	434
Dec	7,640	524
Total	47,339	2,910

The revised gas consumption (for heating only) can now be plotted against the degree days (Figure 3.6).

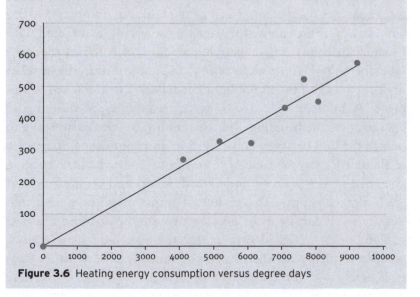

Figure 3.6 Heating energy consumption versus degree days

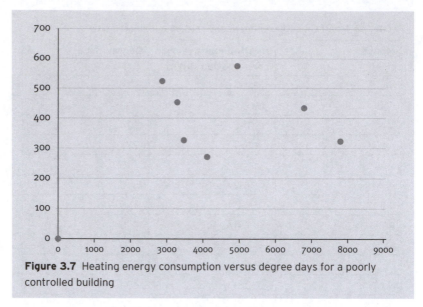

Figure 3.7 Heating energy consumption versus degree days for a poorly controlled building

The regression routines as found on standard spreadsheet packages can be used to obtain the best straight line and the equation of the line. The slope of this graph gives energy use per degree day. Multiplying this by 24 gives W/K, the specific heat loss or total heat loss coefficient (TLC) of the building. In practice there will always be a degree of scatter on the graph, associated with the normal behaviour of people in buildings, such as the periodic opening of windows or doors, and variations in the number of occupants and the use of lights, machinery or electrical equipment. A number of reasons associated with the design and operation of the building and the building services also serve to reduce the link between consumption and conditions. Examples of these are extreme time lags and poor control of the heating system, an over-sized or undersized boiler, and poor control of the boiler air supply. The scatter of points may be so great that that it is difficult or even pointless to attempt to find the best straight line. A high degree of scatter may indicate a serious problem either with the heating system itself or with its controls, such as a broken sensor or actuator (Figure 3.7). These matters should be investigated thoroughly before instigating other energy-saving measures. Problems such as these will also tend to manifest themselves as poor environmental conditions for the occupants – it will be too hot or too cold. Since the main function of the building services is to keep the occupants comfortable, if it fails to achieve that, then the energy used is being wasted.

A range of measures for improving the energy performance of a building is investigated in Chapter 4, where some calculations are presented, and economic assessment is discussed in Chapter 6.

A list of the best options can then be drawn up and presented to management in the form of an energy audit report. The investment appraisal is based on the information provided by the survey appraisal, and determines the final recommendations for improving the building's energy performance.

The energy audit report should include:

1. Description of building – dimensions, materials, location, orientation, purpose (office, workshops, labs, etc.). Hours used.
2. Description of heating/cooling/lighting systems, air-handling units, boilers, fittings, controls.
3. Thermal comfort – state whether conditions are acceptable.
4. Energy consumption – bills, estimates. If not available, state what you would need to do to find out – what measurements to make.
5. Comments on specific points about the operation of the buildings. Does everything work as it should, is it well maintained, and so on?
6. List of where you think energy is being wasted, where and how savings could be made.
7. Calculations based on 6 to show the cost-effectiveness (or otherwise) of measures.
8. Specific recommendations based on 6 and 7,

Check list for energy audits.

Pre-inspection

Location

Building form and orientation

Floor plans and elevations

Drawings of services layout

Areas, floor–ceiling heights

Use

External environment

Maintenance records

Alterations and improvements

Energy sources and tariffs

Determine whether the building is owner-occupied

Determine whether the owner or occupier pays the fuel bills

Determine whether heating/electricity are flat-rate or metered

Fabric survey

Roof – type, condition, insulation, condensation rooflights?

Walls – type solid, cavity material, thickness, insulation, condensation?

Floors – type, insulation?

Doors – draughtstripping, opening type, closers, revolving?

Windows – frames, single/double condition, how well fitting, draughty, openable area and orientation, curtains, blinds coatings daylighting?

Conservatories/atria – orientation type, size, heating or not, what used for, blinds, ventilation?

Energy supply

Electricity

Metered, sub metered, on site generation CHP?

Gas – piped, storage tank

Liquid/solid fuel – storage

Space heating

Energy source – boiler plant type, age, condition efficiency heat recover location and type of emitters, zoning, controls, standard of maintenance?

Check thermal comfort

Hot water – energy source, type, local or central, condition, temperatures, control.

Ventilation – natural or mechanical – fans, size, filtration, condition, ductwork, control, maintenance, heat recovery, recirculation.

Cooling – natural or mechanical, type and size of plant, control, heat recovery?

Lighting – type, installed wattage, age, condition, control, cleanliness.

Computers and other office equipment – type efficiency , energy star-ratings, use.

Appraisal of results of survey

Assess results of upgrading fabric – walls, roofs, floors and windows. Assess cost and energy savings, and other benefits such as improved comfort.

Windows – assess condition of window frames and need for upgrading – this will factor into the case for double glazing.

Assess opportunities for changing window size.

Assesss effect of blinds or shading on cooling load.

Doors – assess opportunities for reducing heat losses – revolving doors, air curtains, closers.

Energy – assess effects of switching fuels, suppliers.

Assess effects of changing boilers and/or heating system.

Consider use of heat pumps/CHP.

Hot water – assess effect of changing to/from point of use heaters.

Ventilation – assess potential for heat recovery, changing to natural ventilation.

Cooling – assess potential for improving efficiency, assess potential for natural cooling.

Lighting – assess potential for improving efficiency with new luminaires or improved controls.

Refurbishment Considerations

Opportunities for instigating energy management procedures do not present themselves continuously. Company policy may dictate that plant may only be replaced when it has come to the end of its useful life, and fabric enhancement may only be considered viable during a major refurbishment. A refurbishment programme offers some of the best opportunities to upgrade the building fabric and install more energy-efficient HVAC and lighting plant, and may be carried out for a number of reasons;

- Change of use of building
- Moving into an old vacant building
- Plant has reached the end of its useful life
- Building fabric is in a rundown condition
- Conditions in building no longer acceptable

The level of refurbishment may include replacing carpets, partitions, light fittings, boilers, heat emitters, entire heating system, ventilation system; removal and replacement of finishes and insulation on outer walls; fitting false ceilings, raised floors, new internal walls; adding air conditioning units; replacing control systems and fitting electronic Building Management Systems (BMS). The lifetime of different elements of the building – structure, services, etc., varies (Table 3.4) and this may dictate the extent of possible changes.

Table 3.4 Lifetimes of different elements of a building

Element	Lifetime (years)
Structure	50 +
Building services	20
Cabling	12
Office equipment	6-8
IT hardware	3-5

CHAPTER 4
TECHNIQUES FOR REDUCING ENERGY CONSUMPTION

The general strategy for reducing energy consumption should be:

- Repair faults
- Reduce loads where possible
- Use efficient plant to service the loads
- Use efficient sources of energy to operate the plant.

Once the sources of energy waste have been identified through the energy audit, techniques for reducing this waste must be found. These can conveniently be placed under the following headings:

- building fabric
- building services plant
- controls
- management of the building
- energy supply.

These are considered in detail in this chapter, with the exception of controls which are covered in Chapter 8.

The availability of finance plays a considerable role in determining the kinds of intervention possible, but practical considerations also dictate to a large extent the kind of changes that can be made.

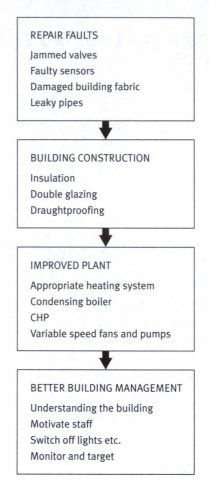

Figure 4.1 Typical techniques to help reduce energy consumption

Repairing faults

Any approach to improving building performance must first ensure that the existing fabric is sound and that plant and controls are operating correctly. Any damage to the building fabric such as broken windows, loose tiles, damp or missing insulation, will compromise the energy performance and must be attended to.

Typical problems include dripping taps, leaking pipes, jammed or broken control valves, broken or inaccurate sensors, and clogged filters. A system of maintenance and repair should be instigated to ensure that faults do not remain unattended for long periods. The use of a building management system (BMS) (see Chapter 8) often proves to be of help in detecting some of these problems. Building management problems which can be solved easily and which will produce quick results include the following.

Items left on overnight such as computers, photocopiers and printers

Simple solutions may have far-reaching consequences – switching off our computers before we leave work can save 70 per cent of their energy consumption. Some energy-consuming appliances such as de-icers for external steps, ramps and car parks may be on without anyone knowing. Escalators which continue running when the building is closed will consume considerable amounts of energy over a year.

Building fabric

Heating, cooling and lighting constitute the major energy users in most buildings, and are largely determined by the basic form and construction of the building, which normally cannot be altered. However, major refurbishment may present opportunities to add a conservatory or atrium, add rooflights or lightshelves to improve daylighting, or even to change the proportion of glazing on the façade. All of these could enhance the building's energy performance. The greatest opportunities for fabric changes lie in adding thermal insulation, replacing single glazing with double glazing, and reducing infiltration by installing draught-proofing materials.

Insulation

Insulation added to the inside of a wall reduces the usable floor space to a small extent, but cavity or externally applied insulation do not. Externally applied thermally insulated cladding is particularly useful for high-rise flats; it is easier to install than internal insulation to each flat, and is cost-effective. External insulation also retains the thermal mass of the building, which helps to mitigate temperature variations. The effect of this thermal mass is lost when internal insulation is used. In houses the easiest location to add insulation is in the loft space, as the process of installation causes little disruption, and a thick layer may be used without intruding on the living space.

Insulation of ground floors in existing buildings is easiest with suspended floors, provided there is sufficient crawl space underneath; there are a variety of arrangements of insulation, utilizing rigid insulation boards, or loose-fill mineral wool supported by netting (Figure 4.2). Insulation of solid floors is more difficult, but strong rigid insulation boards can be laid on top of concrete floors, and wooden flooring laid over the top. This of course raises the floor level, causing significant additional work and costs.

Insulation sometimes fails to deliver the anticipated energy savings. The occupiers may allow internal temperatures to rise, since they can now afford more comfortable conditions; also, after some time loose-fill insulation may settle, leaving uninsulated areas. The effect of studding to support the insulation should is sometimes ignored (Figure 4.3) – it has a higher thermal conductivity and increases the effective U-value, but if the correct U-value calculation is used, this will be taken into account.

Cavity insulation may fill all or part of the cavity (Figure 4.4), and produces different U-values. A 100 mm cavity with partial fill of 50 mm insulation has a thermal resistance value of 1.38 m^2K/W, while filling the cavity completely with insulation would produce a value of 2.5 m^2K/W. Normally in a retrofit beads are blown into the cavity to produce full cavity insulation.

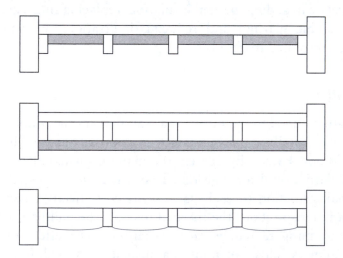

Figure 4.2 Insulating below suspended floors

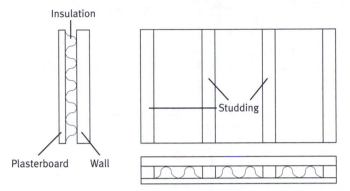

Figure 4.3 Internal insulation supported by studding. The studding has a higher conductivity than the insulation, resulting in a higher overall U-value.

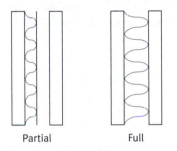

Partial   Full

Figure 4.4 Partial and full cavity wall insulation

Thermal bridging

In all building elements there may be thermal bridges: that is, areas of low thermal resistance connecting the inner and outer surfaces, through which heat passes relatively easily. These can significantly reduce the effectiveness of insulation in comparison with the estimated reductions in energy consumption. Such bridges typically occur at corners, window cills and soffits. Careful detail design is needed in order to avoid these (Figures 4.5 and 4.6), but in many instances the basic building structure means that they are unavoidable.

The thermal bridge through the window frame can be avoided by using frames that have a thermal break, although the cill and soffit may still constitute a thermal bridge.

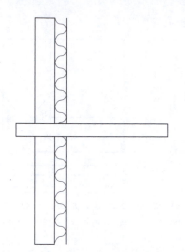

Figure 4.5 Thermal bridge at a balcony. Heat flowing out of the building can bypass the wall insulation unless the floors and ceilings are insulated.

Figure 4.6 Thermal bridge at a window

Reducing cooling load

Cooling loads in office buildings have increased greatly in recent years to due to the almost universal use of computers. The additional loads may amount to 10 W/m^2 or more. Reducing the solar gains by the use of internal blinds and shades is a cost-effective way to reduce the overall cooling load, and may help to reduce the peak gain in a south-facing room by 15–20 W/m^2. At a higher level of investment, external shading, fixed or movable, can be even more effective.

Ventilation and draughts

Reducing draughts is another cost-effective technique and can be achieved in a number of ways. In dwellings the application of draught-proofing material round the edges of doors and windows is usually sufficient, while in commercial premises, the use of revolving doors and automatic door closers should also be considered.

Calculations showing the effects of insulating buildings, and other interventions are given in Case Studies 1 and 2 and Appendix 1.

Building services equipment

The building services equipment uses most of the energy consumed in a building. For maximum energy efficiency it should be appropriate to its function, operate efficiently, and be correctly sized. Many systems are significantly oversized and therefore run at much less than optimum efficiency. It is worthwhile investing in smaller or more efficient items of plant where feasible, as small changes in the rating of a piece of equipment may lead to large reductions in lifetime energy use. For instance, an 11 kW motor costing £700 can consume £67,000 worth of electricity over its lifetime of ten years, and a small increase in capital cost to acquire a machine of greater efficiency will pay for itself very quickly.

Heating system

Choosing the right heating system has important energy implications. Some systems such as warm-air heating have an output that is almost totally convective, while others are predominantly radiative. Radiators used in central heating systems emit between 30 and 70 per cent of their energy by radiation, depending on the design and the surface temperature. Older systems often have large cast-iron radiators which contain a large volume of water, have a high thermal mass and therefore a slow response. More modern radiators are thinner and contain less water, therefore they are capable of more rapid response to conditions and provide greater efficiency.

In buildings with high ceilings, a warm-air system may turn out to be very inefficient, since the warm air will rise towards the roof. Destratification fans and ductwork can be fitted to direct the warm air back to ground level and reduce temperature gradients, improving efficiency. Often such spaces are better heated by high-level radiant heaters, which beam radiant heat directly to the occupants and surfaces. Intermittently heated spaces with large amounts of thermal mass may also benefit from high-level radiant heating which heats the occupants directly. An efficient alternative is underfloor heating, which requires lower water temperatures and can be used effectively with efficient sources of heat such as heat pumps and condensing boilers.

Boilers

Modern conventional gas and oil-burning boilers for heating systems are very efficient, and can reach a maximum efficiency of 75–85 per cent at or near maximum capacity. At lower load fractions efficiency may drop significantly, because of the standing losses. When a large heat load needs to be met, it is usually more efficient to use a number of small boilers, since one small boiler at or near maximum efficiency is more efficient than a large boiler at low load.

Condensing boilers

Although conventional boilers are now very efficient, even greater efficiency can be achieved by using condensing boilers. When a fuel such as natural gas is burnt, it produces carbon dioxide (CO_2) and water vapour.

$$CH_4 + 2O_2 \rightarrow CO_2 + 2H_2O + \text{water vapour}$$

Since the water is in the form of vapour at an elevated temperature, the heat content is significant, and if it is allowed simply to disappear up the flue then it is lost. The latent heat and some additional sensible heat can be recovered by using condensing boilers, which are available from domestic scale upwards. After combustion and passing over the main water tubes, the flue gases pass over further heat exchange surfaces where the water vapour condenses out (see Figure 4.7). Near the top end of the load cycle, the efficiency of such boilers is about 10–15 per cent greater than for a conventional boiler, and even when not operating in the condensing mode, some advantage is obtained due to the greater heat transfer surface. Condensation can only take place when the return water is at a temperature less than about 59°C. The lower the return water temperature, the greater the degree of condensation and the higher the efficiency of the boiler.

Condensing is particularly effective for a system with a low water return temperature such as underfloor heating. A 250 kW boiler with a 40°C return water temperature produces 14 l/hr condensate, therefore a drain is required to remove it. The condensate is slightly acidic, therefore the drain should be of plastic, and stainless steel flues are preferred since copper and cast iron are susceptible to corrosion. A fan is normally required to remove the flue gases, which are cooler (at 55–100°C) and therefore of lower buoyancy than in a conventional boiler. The cost is 30–50 per cent higher than a conventional boiler and the payback period is two to five years.

Table 4.1 Efficiency of a condensing boiler at a range of return temperatures

Mode	Efficiency %
Non-condensing	85
Return temp 40°C	90
Return temp 30°C	95

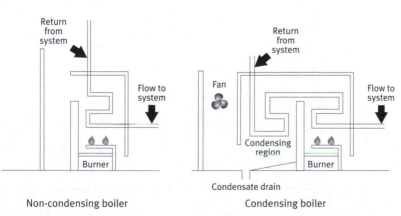

Figure 4.7 Conventional and condensing boilers

Variable speed drives (VSD) for pumps and fans

Pumps consume about 12 per cent of the world's electrical energy, and most of them consume over 50 per cent more than they should if designed, specified and used correctly. The energy use over its life cycle is many times the cost of a pump, so it is worthwhile investing in more efficient pumps and control devices. Older heating, ventilation and air-conditioning (HVAC) systems often have only single-speed fans and pumps, while newer ones may include two-or three-speed plant. Since pumps and fans need to be sized to meet the maximum demand, for a substantial proportion of their life the load is only a fraction of this and energy is wasted. Figure 4.8 shows the operation of a fan in a ventilation system. Where the fan curve and the system curve cross (A) is the operating point, and the power consumed by the fan is given by the area ADOG. When demand falls the airflow rate is reduced by closing a damper, resulting in an increased pressure loss and reduced flow, and shifting the operating point to B.

The power consumption is now given by the area of rectangle BEOH, which is little lower than the original value, although the flow rate has fallen by almost 50 per cent. An alternative way of restricting the flow is to use an inverter drive, which allows the flow rate to be reduced by slowing down the fan, resulting in a corresponding lowering of the energy input; thus the power consumption at flow rate E is given by the area CEOF, around one quarter of the value obtained by using dampers. Annual fan and pump energy savings of over 60 per cent are claimed by users of VSDs, which in many cases can be retrofitted to existing systems.

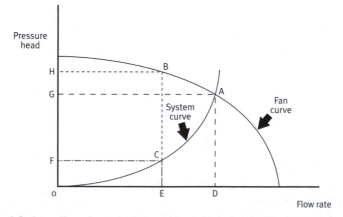

Figure 4.8 Operation of a variable speed fan with inverter drive

Domestic hot water

In homes, hot water can represent up to 25 per cent of the overall energy demand, and almost half of the heating requirement; in a well-insulated building it may even constitute the dominant heat load. In offices the proportion is lower, but when refurbishing it may often be more efficient to use point-of-use water heaters, even if electrically heated, as this avoids the need to use the boiler during the summer months.

Heat pumps

A heat pump used in heating mode upgrades low-level heat by extracting it from a source such as the air, the ground or running water, and releasing it at a higher temperature to provide space heating in a building. It employs a compression–refrigeration cycle and a pump which is usually electrically driven, and is therefore useful where there is no mains gas supply. Heat pumps are supported by the British government under the renewable heat incentive (RHI). Their advantages include reduced servicing, no boiler, high security, long life, no flue or ventilation and no local pollution. In appropriate situations they can save money and reduce carbon emissions.

The efficiency of the system is defined by the coefficient of performance (COP), which may have different values for heating (COP$_h$) and cooling (COP$_c$). Typical values range from 3–4 for COP$_h$, and 2.8–3.6 for COP$_c$. They are defined as:

$$COP_h = \frac{\text{Evaporator duty}}{\text{Compressor power}}$$

$$COP_c = \frac{\text{Condenser duty}}{\text{Compressor power}}$$

The COP falls as the temperature difference between the evaporator and condenser increases. At large temperature differences supplementary heating may be needed, adding to the capital costs.

Heating mode

Air-source heat pump (ASHP)

Air-source heat pumps have the advantage of relative simplicity; a fan blows air over the evaporator coil located on an outside wall of the building (see Figure 4.9). Installation of ASHP is relatively cheap, but they suffer from the drawback that the temperature of the air is at its lowest when the greatest amount of heat is required, leading to low values of COP at high load. Also, as heat is extracted from the external coil, regular defrosting is necessary. ASHPs are therefore of most benefit in moderate climates. Typically an air source heat pump can provide 85 per cent of the heat for a building, and a conventional boiler will provide the rest. ASHPs have a typical COP of 4.4 at an air temperature of 7°C and a flow temperature of 35°C.

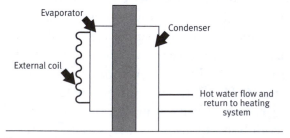

Figure 4.9 Air-source heat pump

Use of water as heat source

Closed-loop system

Because of its high thermal capacity, the temperature of a body of water varies much less than that of the air. As a result, the severe falling-off in COP experienced in air-source systems in severe conditions is much reduced. Sea water in estuaries and water in rivers has been used successfully in this way. In these systems water is used as the circulating heat transfer medium, passing through a heat exchanger on the evaporator side and a coil placed in the water.

Open-loop system

An alternative to a circulating system is an open-loop system which uses water that is pumped from a deep borehole and passes over the evaporator coil, after which it is led to a soakaway. The temperature of the water is close to that of the ground, which in the United Kingdom is between about 9 and 12°C; it is therefore often used for low-cost cooling (Figure 4.10). The running cost of this is considerably lower since the only energy input is that required to pump the water from the ground. The effective coefficient of system performance (COSP) lies between 10 and 100.

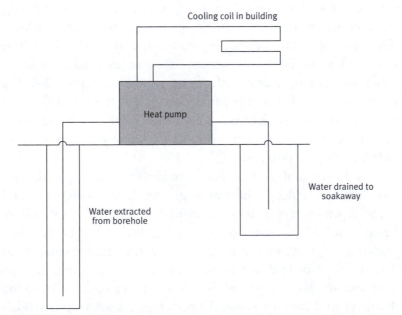

Figure 4.10 Once-through system

A licence to extract water is normally required (from the Environmental Protection Agency (EPA) in England and Wales or the Scottish Environmental Protection Agency (SEPA) in Scotland). The agency may also insist on additional requirements such as returning the water to the aquifer, and there are also charges (currently £10–25/litre in the United Kingdom), which may affect the feasibility of a scheme. Water quality should be high, with low dissolved solids, and available flow rate should be 25–50 l/s. The additional capital costs may be considerable, and include the drilling of test boreholes.

Some notable large developments have used groundwater in association with heat pumps or other systems, including Portcullis House in London, an office block for MPs. The Scottish Parliament building in Edinburgh uses water from a spring for 'free' cooling and expels the water to ponds in the grounds.

Ground source heat pumps (GSHP)

The ground a metre or more below the surface of the United Kingdom has a relatively stable average temperature, ranging from 9°C in the far north to 12°C in the south, while in hot climates such as India and the Far East it may reach 20–30°C. A circulating coil is buried in the ground, and heat exchange between the ground and the water in the coil reaches around 5W/mK. Two approaches may be used, a vertical coil or horizontal. The vertical coil system is more expensive, since to achieve sufficient heat transfer a deep hole, perhaps 80–150 m deep, is required. It tends to be used when the ground area is insufficient for a horizontal coil. A high COP can be achieved, but at an increased capital cost, and cooling capacity is about 500 W/m² of ground area.

A horizontal coil is more common, and typically uses trenches 1.5 m deep, the optimum depth in terms of overall costs. A much larger area of ground is required, 200–300 m² being required in order to extract 10,000 kWh/year at a heat exchange capacity of about 30 W/m² ground area (Figure 4.11). The coils can be laid in a number of ways: single pipe, multiple pipe and spirals. The spiral or 'Slinky' pipes have coils 60 mm diameter and require about 30 per cent less area than straight

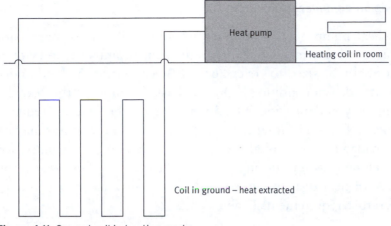

Figure 4.11 Ground coil in heating mode

pipes, but a greater overall length of pipe is required for the same cooling capacity. Where single pipes are used they should be at least 0.3 m apart, and trenches should be separated by 2 m.

GSHPs operate over a range of water input temperatures, with typical evaporator temperatures between −5 and + 12°C while the condenser temperatures range from 50–55°C maximum. This temperature is too low for a conventional radiator system but may be used with warm air or underfloor heating systems. It is also too low for domestic hot water, which must be at a minimum of 60°C to reduce the risk of legionella; a supplementary heater, usually electric, can be used after pre-heating the water to 50°C with the heat pump. Problems may arise if the fluid in the ground loop is below 0°C, as freezing may cause ground heave. Typical COP for a GSHP in UK conditions are shown in Table 4.2.

COPs of 3.5 and over give a significant advantage over gas-fired heating. On average a GSHP produces 65 per cent less CO_2 than oil-fired heating, and 45 per cent less than gas.

Table 4.2 Typical heating COP for a range of temperature outputs

Water temperature°C	COP
55	2.4
45	3.2
35	4.0

techniques for reducing energy consumption **53**

Design considerations

A heat pump may be designed to cover full load or part load. Part-load systems significantly reduce the capital cost and would normally be sized to cover about 50 per cent of the load under winter design conditions. On this basis for most of the year the pump would provide 90–95 per cent of the heating requirements. Top-up heating to cover the lowest temperatures may be provided by low-capital-cost electric heaters or, particularly for dwellings, wood-burning stoves. The source of heat and the type of heat distribution should be matched according to temperature requirements (Table 4.3).

Domestic heat pump use

Although heat pumps are becoming more popular in the United Kingdom, their use is not as widespread as in Scandinavia and some other regions. Packaged units which combine an air-source heat pump and top-up heating, together with controls, are available, and may also include domestic hot water and cooling options. Such systems may also be used in combination with fossil-fuel boilers (connected in series to the heat pump) where the heat pump provides pre-heated water to the boiler for the coldest periods. For refurbishment projects the boiler may also be connected in parallel with a heat pump of lower capacity so that the boiler covers the full load at extreme temperatures. Whether the fossil-fuel boiler or electric heat pump is the most cost-efficient supplier of heat is determined by the relative fuel costs. Such a system provides the flexibility to use the cheapest fuel option at the time.

Table 4.3 Temperatures required for different heat distribution systems

Heating system type	Temperature°C
Underfloor	30-45
Low-temperature radiators	45-55
Radiators	60-90
Warm-air	30-50

Controls for heat pumps

The most energy-efficient form of control uses a variable speed drive to the compressor via an inverter. In this way the output of the pump can be matched to the demands of the building, and as the required heat output falls, so does the energy input to the compressor.

Performance comparison

A comparison between the performance of different technologies is shown in Tables 4.4 and 4.5 for heating and cooling respectively. Generally, heating is cheaper and results in fewer CO_2 emissions per kWh than cooling.

Table 4.4 Comparison of heating performance of condensing boiler, air source heat pump and ground source heat pump

	Condensing boiler	Air source heat pump	Ground source heat pump
Heating load (kW)	40	40	40
Seasonal efficiency (%)	88	300	400
Energy input (kW)	45.4	13	10
Cost of energy (p/kWh)	4.5	11.0	11.0
Hours run	2000	2000	2000
Cost of input energy (£)	4086	2860	2200
CO_2 emitted (kg)	21.6	14.2	10.75

Table 4.5. Comparison of cooling performance of air source heat pump and ground source heat pump

	Air source heat pump	Ground source heat pump
Cooling load (kW)	40	40
Seasonal efficiency (%)	280	320
Energy input (kW)	35.7	31.25
Cost of energy (p/kWh)	11.0	11.0
Hours run	500	500
Cost of input energy (£)	1963	1718
CO_2 emitted (kg)	15.3	13.2

Ventilation and waste heat recovery

Natural ventilation is obviously the lowest energy-cost option, but in cases where mechanical ventilation is unavoidable, fan motors should be fitted with VSD as described earlier in the chapter.

In housing we try to minimize the ventilation and infiltration rates in winter in order to keep heating loads down. However, when a high ventilation rate is essential, as in some light industrial processing applications, then energy costs can be kept low by recovering heat from the waste ventilation air. Devices used for this purpose include plate heat exchangers, run-around coils, and thermal wheels. In ventilation applications air–air plate heat exchangers are frequently used where the inlet and outlets may come close to each other, and thermal wheels have been used successfully in many applications. Plate heat exchangers and thermal wheels have an effectiveness between 70 and 80 per cent, and run around coils up to 85 per cent. Pressure drops within the heat exchangers should not be ignored, and may reach 100–200 Pa in a thermal wheel and up to 250 Pa in run-around coils. Where there is a substantial distance between inlet and outlet, run-around coils are more appropriate. Mechanical ventilation with heat recovery (MVHR) has been used successfully in 'problem' housing to reduce excessive moisture, combating condensation and mould growth, thus saving money on maintenance and decoration in the long term. The energy costs of ventilation can be reduced by using demand-controlled ventilation, based on room CO_2 levels, combined with the use of inverters to control variable-speed fans.

Combined heat and control (CHP)

A CHP system is a heat engine that produces electricity through a generator and is able to make use of the rejected heat from the engine in a water or space-heating application. Many small CHP units use diesel or gas power and have similar reliability factors to conventional boilers; availability factors of 95 per cent and over are common. The great advantage of CHP is that overall energy efficiency is in the region of 80–90 per cent, as shown in Figure 4.12. A typical energy balance is shown.

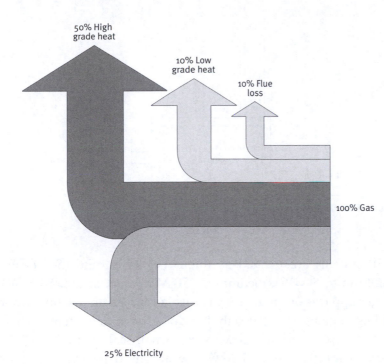

50% High
grade heat

10% Low
grade heat

10% Flue
loss

100% Gas

25% Electricity

Figure 4.12 Outputs from a combined heat and power (CHP) system

Low-grade heat refers to heat emitted at 35–45°C, while high-grade heat has temperatures in the region 70–85°C. The viability of a CHP system depends on a number of variables: a simultaneous demand for heat and electricity, the ratio of heat to electrical load, the capital cost, the fuel costs for the system, and the cost of the alternative (usually mains) electricity and heat. Viability depends to a large extent on the 'spark spread', that is, the difference in utility rates. To make CHP worthwhile the cost of electricity should be three times the cost of heat.

General CHP considerations

In general, a CHP system needs to operate for at least 4000 hours per year to be cost-effective. If well designed, it can give a payback period (PBP) of three to five years.

Figure 4.13 shows a typical CHP system which consists of an engine or turbine, a generator, a heat recovery system, an exhaust system and a control system.

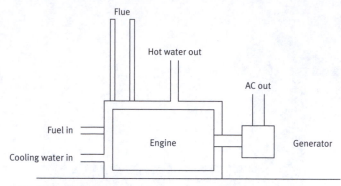

Figure 4.13 CHP schematic

CHP systems are defined as micro (<5 kW), mini (5–30 kW), small (up to 5 MW), medium (5–50 MW) and large (>50 MW). Larger applications may use gas turbines or steam systems, while Stirling engines are used only for domestic-scale systems, where electricity production is low. Many small-scale units having an electrical output up to 1 MW use automotive-derived engines altered to run on gas.

The heat to power ratio is typically as follows:

up to 3:1 microturbines
up to 2:1 engines
1:1 fuel cells.

Dual-fuel engines tend not to be used as they are not optimized for efficiency for either fuel, therefore their potential CHP efficiency is low.

Most of the heat (50 per cent) is recovered from the engine cooling jacket and the exhaust system in the form of low-temperature hot water (LTHW), with low-grade heat being the remainder (10 per cent). The heat from the exhaust gases is recovered using a gas-to-water heat exchanger with the heating circuit water flowing directly through it. A further condensing heat exchanger may be added to remove latent heat from the flue gases. A buffer tank may be installed between the CHP and the hot water return to reduce frequent cycling.

As connections to the mains will normally be maintained, agreements have to be made with the local distribution network organization. If the electricity generation is less than 16A per phase a less formal arrangement is required but the distributor must be informed.

Criteria for installing CHP

A reasonably large heat load should exist over the greater proportion of the year, to enable the plant to run for 11 hours per day for the whole year, or 17 hours per day for eight months of the year. Heat and electrical loads should be simultaneous, space should be available for the unit, and the electrical load should remain above the output of the unit for most of the operating hours. Mini CHP should be sized to match the base heating load to maximize running hours, and in a mixed system the CHP should be the lead heating appliance. If the CHP is replacing gas for heating, then any gas used for cooling should be discounted, as it cannot be replaced by CHP heat. The overall fuel efficiency should be in excess of 75 per cent.

In order to qualify for government incentives such as a Climate Change Levy (CCL) exemption under enhanced capital allowances, the CHP needs to meet a certain quality standard. A Quality Index (QI) is defined as:

$$QI = 249 \eta power + 115 \eta heat$$

where power efficiency $\eta power$ = total power output (MWhe)/ total fuel input(MWh)

Heat efficiency ($\eta heat$) = Qualifying heat output (MWTh)/ total fuel input (MWh)

The constants 249 and 115 are related to the alternative electricity supply and alternative heat supply options displaced by the CHP. Qualifying heat output is the amount of useful heat supplied annually from a scheme that can be directly shown to displace heat that would otherwise be supplied from other sources. QI should be at least 105 with a minimum power efficiency of 28 per cent to qualify, and metering is required for input and both outputs.

Example of CHP calculation for residences at a university campus in the United Kingdom

In this example heating is provided by gas and cheap off-peak electricity is available at night. It is generally not cost-effective to run the unit during the night because of this lower cost. For the campus, mains natural gas is the obvious fuel for the CHP as it is already available at a favourable tariff.

Current energy use for residences is:

	Consumption kWh	Cost £
Electricity @ 11p/kWh	874,564	£ 96,202
Gas @ 4.5p/kWh	2,679,311	£120,568

taking existing boiler efficiency as 80 per cent and with an estimated load factor of 75 per cent.

For year-round base load operation, we need the hourly useful heat load (H_L) averaged over the summer months.

$$H_L \text{ (kWh)} = \frac{(H_{June} + H_{July} + H_{August}) \times \text{boiler efficiency}}{(\text{No of days x Hours per day})}$$

$$H_L = \frac{(116728 + 140188 + 120036) \times 0.8}{(89 \times 20)}$$

$$H_L = 169 \text{ kW.}$$

$$E_L = \frac{(E_{June} + E_{July} + E_{August}) \times F}{(89 \times 20)}$$

Where F is utilization and is 1 for separate day and night metering, 0.85 for flat rate metering.

$$EL = \frac{(53832 + 58335 + 50026) \times F}{(1780)}$$

$$= 77.45 \text{ kW.}$$

Therefore average daytime low electrical demand is 77.45 kW. If no electricity was exported, then CHP would need to have an electrical output of no more than 77.45 kW. The electricity cost for this period = £17841 and gas cost = £16,962.

Selecting from a manufacturer's table of outputs, unit A would be suitable, with an electrical output of 77 kWe, a corresponding heat output of 123 kW and a gas input of 233 kW. Savings are worked out as follows:

Unit A

Savings	p/hour	Total p/hour
Electricity 75 × 11 Gas 125/0.8 × 4.5	825 703	1528
Costs		
CHP gas 233 × 4.5 Maintenance @ 0.7p/kWh$_e$ Electricity generated = 0.75 × 75	1048.5 52.5	1101
Net benefit		427

The saving is £4.27 for every hour of CHP operation. Therefore the annual number of operating hours can be given by the load factor × total hours per year = 0.8 × 8760 = 7008 hours, minus 5 days per year for servicing = 6888 hours. Therefore the potential hours of operation are 6888, multiplied by 4.27 gives annual savings of £29,411.

Since the electricity could be exported, a base heat load sizing method could be used. Unit B, with an electrical output of 85 kW and a corresponding heat output of 143 kW, and gas input of 254 kW could be used.

Unit B

Savings	P/hour	Total p/hour
Electricity 85 × 11 Gas 143/0.8 × 4.5	935 804	1739
Costs		
CHP gas 254 × 4.5 Maintenance @ 0.7p/kWh Electricity generated = 0.75 × 85	1143 63.75	1206.75
Net benefit		533.75p/hour

Net saving is £5.33 for every hour of CHP operation, and the annual potential savings are £36713.

The CO_2 savings from such a scheme can also be estimated. The CO_2 saved per hour from electricity generation by the CHP unit as opposed to buying it from the grid equals 70.55 kg. There is, however, an increase in the use of gas. The existing CO_2 emission for gas equals 35.75 kg/hour, and the new CO_2 emissions are 50.8 kg/hour. Therefore there is an increase of 15.05 kg/hour. The net benefit is therefore 70.55 - 15.05 = 55.5 kg of CO_2 saved per hour. This results in an annual saving of 382 tonnes of CO_2 per year.

In this case there is no demand for cooling, as it is not justified by the climate.

Combined heat and power summary

The general design procedure is to identify the heat demand profile and the electricity demand profile and size according to the heat load. It is best to avoid heat dumping by use of thermal storage, which enables electricity sales to be maximized. The inclusion of a large number of diverse users affords opportunities to even out loads over time, thus making for greater efficiency.

Air conditioning

Where there is an existing air conditioning (a/c) system significant improvements may be difficult to effect, but regular servicing and maintenance should be implemented to achieve optimum performance. Many a/c plants are oversized to cope with the hottest conceivable day, with some capacity to spare; as a result, compressors run most of their lives well below full load and with low efficiency. A refurbishment may offer opportunities to replace some items of plant with smaller, more efficient units. Changes to the building fabric and operation may reduce some of the plant loads significantly to allow this.

Night ventilation to reduce the daytime cooling load by reducing building fabric temperatures is often incorporated into the design strategy of new buildings, and may be implemented in an existing building. The feasibility of the doing so will depend on a number of factors, such as the structure and layout of the building, the expected outdoor temperatures, and the daytime heat gains in the building. In a heavyweight building where night temperatures fall well below 20°C and there is a cooling load during the day, then night cooling should be seriously considered. It is essential that there are sufficient internal openings to allow the ventilation air to flood the building, but security should not be compromised. If daytime temperatures exceed 36°C then night ventilation is unlikely to have a significant effect.

Among the possibilities for improving energy efficiency is the replacement of mechanical ventilation and a/c with natural ventilation. To assess the viability for natural ventilation the following considerations should be borne in mind.

Environmental

Peak summer heat gains – if they are in excess of 40–50 W/m^2 then mechanical ventilation or air conditioning are likely to be necessary. Noise from roads or railways entering the building may be a serious problem with a naturally ventilated building. The noise level should be at most 70 dB at the façade to achieve an acceptable level of 50–55 dB with windows open. Airborne pollution from traffic on roads may be unacceptable. The presence of a prevailing wind should be noted as it may affect the strategy adopted. The effect of surrounding buildings, and on them, with respect to light, wind movement and sound, should be considered.

Building

Shallow-plan buildings are easier to naturally ventilate than deep-plan. For single-sided the maximum depth is 7–10 m, for double-sided cross ventilation (open plan) up to 15 m is possible.

The specific fan power (SFP) indicates the number of watts required to move 1 litre of air per second, and for new fans in ventilation and a/c systems should be as low as possible. The current maximum allowed is 0.8 W/l/s and it is likely that the Building Regulations will require a figure of 0.6 in future. Electrically commutated (EC) motors are more efficient, and can give a SFP of 0.3 compared with 0.8 in a conventional motor, and speed can be controlled better and more efficiently.

Lighting

Lighting accounts for a significant proportion of energy costs in commercial buildings and therefore presents huge opportunities for savings. Although heating is the major user of fuel, because of the cost differential between heating fuels and electricity, lighting may represent 30–50 per cent of the total energy bill.

When a room is too warm we notice very quickly, and make adjustments to the temperature, but we do not necessarily notice when there is too much light, as the eye can adapt over a wide range of lighting levels. Because of this, lighting is often poorly controlled. Energy can be saved in lighting in a number of ways – by increasing its efficiency, by reducing the level of lighting, or by cutting the hours of use.

Lighting control options

Manual control

- Position switches so as to encourage users to switch off.
- Group luminaires and switches to take advantage of day-light near windows.
- Colour-code groups of lights in buildings such as super-markets for different sets of users who require different lighting levels, such as cleaners, shelf-stackers and customers.

Timed control

- Switch lights off at natural break times and at the end of the working day. Allow manual override of main lights or use locally controlled task lighting.

Sensor control

- Use occupancy-detection sensors (such as passive infra-red (PIR) or ultrasonic).
- Use daylight sensors to switch groups of luminaires off when daylighting is adequate.
- Use daylight sensors in conjunction with dimmers to use electric lighting to top up daylighting.

- Use individually addressable luminaires for greater flexibility.

Escalators and lifts

In certain building types such as department stores and shopping malls, escalators may consume up to 15 per cent of the total energy used in the building. In such buildings they are often running continually during opening hours, and in some cases considerably beyond. The first priority should be to ensure that they do not run beyond the hours required; this is easily achievable with a time switch. Further savings can be achieved by more sophisticated control. For instance when there is no demand the speed can be reduced, or the escalator stopped altogether. Speed reduction can be achieved using star-delta control or variable voltage variable frequency drive (VVVFD). The greatest energy savings are achieved by stop/start control followed by VVVFD, but it results in more wear to the moving parts and increased maintenance costs. Lifts can also benefit from VVVFD which enables energy consumption to be reduced through speed reduction.

Use of renewable energy in buildings

A good general principle to follow is that the load should first be reduced if possible, then means of meeting the load found, whether that is from a fossil or renewable source. There are opportunities to use renewable energy in buildings to reduce reliance on fossil fuels. The feasibility of the different technologies is a function of climate, capital cost, and the cost of fuel replaced, bearing in mind that for many systems such as photovoltaics (PV) and wind turbines it is necessary to maintain a mains connection and add a control unit. Depending on location and local economics and fuel costs, some of the following may be appropriate.

Possible systems include solar water/air heating, solar electricity (PV) and building-integrated PV, solar refrigeration, and roof-mounted wind turbines. It is considered by many that micro wind turbines are unlikely to pay back either their carbon emission or costs, and that larger turbines sited more remotely, which are more efficient, are the future for wind power. Transmission losses from such locations are not great – only 10 per cent losses are incurred by transmitting the whole length of the United Kingdom. PV are not so sensitive to scale factors, but it remains the case that the control gear for a very small PV system costs not much less than that for a large one.

Summary

A number of measures can be implemented to improve any building. The cost-effectiveness of each measure will vary with location and with the individual building. A rough order of effectiveness is given in Figure 4.14 (in approximate order of decreasing cost effectiveness based on cost per kg CO_2 saved). The measures applicable to specific building types will vary greatly; examples of potential savings in shopping centres are shown in Table 4.6.

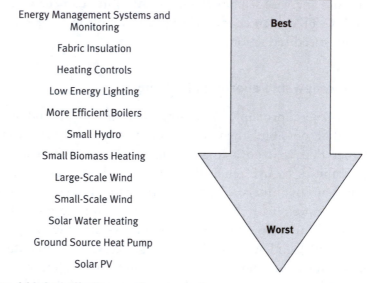

Energy Management Systems and Monitoring

Fabric Insulation

Heating Controls

Low Energy Lighting

More Efficient Boilers

Small Hydro

Small Biomass Heating

Large-Scale Wind

Small-Scale Wind

Solar Water Heating

Ground Source Heat Pump

Solar PV

Best

Worst

Figure 4.14 Cost-effectiveness of energy saving measures

Table 4.6 Potential energy savings in retail shopping centres

Measure	Potential saving
Switch off unessential lights outside opening hours. Use timers to switch off most lights when centre closes	10%
Switch off unessential lights in back-of-house areas	5%
Replace tungsten bulbs with compact fluorescent	75%
Install photocell controls to make use of daylight. Use presence detectors in areas such as toilets	20%
Switch off escalators outside running hours	15%
Switch off external and car park lighting when adequate daylight	60%
Optimize switch-off of HVAC system	20%
Reduce air conditioning by night cooling	20%
Set thermostats correctly	1ºC = 7%
Check control strategy so not heating and cooling at same time	10%

CHAPTER 5
INSTRUMENTATION AND MEASUREMENTS

An old adage of management is 'if you can't measure it, you can't manage it', and while information from the energy bills can be used in the energy audit, it is often necessary to have additional information taken from measurements of the conditions in the building, the state of the plant, and the energy consumption. Additional measurements can be carried out to evaluate the performance of plant items. Monitoring of a building over a short or an extended period may be considered necessary in order to elicit this information. Equipment for this purpose includes data loggers and sensors. With the right equipment, the measurements required for energy management can be made relatively easily, and modern electronic instruments are compact, durable, and inexpensive.

The main measurements of interest are:

- fuel consumption
- temperature
- electrical power
- air movement
- water flow
- relative humidity
- heating efficiency
- U-value.

Fuel consumption

This refers to the overall consumption of cubic metres of gas, litres of fuel oil, kWh of electricity, and so on. Conversion to standard units of kWh and carbon dioxide (CO_2) emissions may be carried out using Table A5 in Appendix 2. For meaningful comparison with other buildings of the same type, the consumption should be calculated on the basis of square metres of treated floor area. Used in conjunction with internal and ambient temperature readings, they can produce degree day information for the kind of analysis which was shown in Chapter 2. Where a building management system (BMS) is used, it can be configured to collect and store meter data.

Temperature

The most important measurements, and the most numerous in building applications, are measurements of temperature. Typical measurements required are internal and external air temperature, radiant and globe temperature, surface temperatures and fluid temperatures in pipes and ducts. The range of sensors used with data loggers and BMS includes those with an electrical output such as thermocouples, thermistors and resistance thermometers. The temperature measured by a globe thermometer is generally considered to give a response closely related to that of humans, and combines the effects of air and radiant temperatures. The instrument comprises a temperature sensor located at the centre of a 40 mm diameter sphere whose surface is painted matt black. The operative temperature of a space (see Appendix, page 158) may be assessed by taking the average of the globe temperature measured in a number of places.

All bodies at temperatures above absolute zero emit infrared (I-R) radiation in wavelength region 0.1 to 100Ω, and the temperature of a body can be measured as a function of the thermal radiation emitted by it. I-R thermometers allow the surface temperature of a body to be measured using a non-contact method, which has the advantage that measurements may be made from some metres away from the surface. I-R cameras are regularly used in thermographic surveys, in which they produce an image of the object being investigated, colour-coded according to temperature. By surveying the outside of a building, preferably at night and in winter, it is possible to locate places where excessive heat is being lost, for instance where insulation is missing or where there is a thermal bridge. Small hand-held I-R cameras are now available inexpensively for purchase or for hire. Considerable expertise is needed to avoid misinterpretation of the results, however.

A number of inexpensive hand-held instruments are available for measuring temperatures; they are useful for spot measurements of conditions, and some have outputs to data loggers.

Electrical power

Electrical energy is used for lighting, small power-load equipment (such as personal computers (PCs), copiers and printers), fans, pumps, immersion heaters, lifts and escalators, and other factory equipment such as machine tools and compressed air machines.

While the overall electricity consumption is usually metered, it is unusual to find separate metering to individual circuits. To isolate a circuit or individual item of plant or appliance, a current clamp is used; the current induced in the clamp is proportional to that in the primary circuit and can be fed to a data logger. For items drawing constant power, hours-run meters can be used, although they give no information on the timing of the power consumption.

Air movement

Measurements of air movement may be made for a number of reasons. They include:

- local air movement in rooms
- movement of air through ducts
- ventilation or infiltration rates

Measurement of local air movement

Qualitative assessments of air movement, draughts or leaks can be made using plumbers' smoke pellets: local air flows within rooms are usually too low to make any accurate measurements of air velocity.

Movement of air through ducts

Ventilation rates in mechanically ventilated buildings can be assessed by measuring the rate of airflow through the ducts. The range of instruments include:

Rotating vane anemometer

These are simple and relatively robust; a small vane assembly is coupled to a mechanical counter via gears. As this is a mechanical device the lower limit of measurement (0.5 m/s) is determined by friction in the bearings and gears.

Hot wire anemometer

The sensing head comprises a wire which is heated by an electric current, and when placed in the air stream a cooling effect is exerted on the wire, by convection. As the air speed increases, the convection heat transfer coefficient increases, cooling the wire more. The filament is connected to a bridge circuit, and the device gives a readout in velocity, or a voltage output to a logger. Compensation for the air temperature is included. As there are no moving parts to cause friction, velocities as low as 0.4 m/s can be measured.

Pitot tubes

Pitot-static tubes comprise two tubes, one facing the air flow directly and the other perpendicular to the flow: they are normally contained in one measuring head (Figure 5.1). When air enters the tube facing the air stream, its kinetic energy is converted to pressure energy, therefore the pressure in that tube is equal to the static pressure in the duct plus the velocity pressure. The static pressure is measured by the tube perpendicular to the flow. The air is initially at velocity V which reduces to zero as it is brought to rest in the tube. The pressure differential ΔP is equal to the velocity pressure and is given by:

$$\Delta P = 0.5 \, \rho_a V^2$$

Where ρ_a is the density of the air (normally taken as 1.2kg/m³).

$$V = (1.66\Delta P)^{0.5}$$

If ΔP is measured in a liquid manometer,

$$\Delta P = \rho_l gh$$

Where ρ_l is the density of the liquid in the manometer, h is the differential in the column heights, and g is the acceleration due to gravity (9.81m/s²).

$$V = (16.35\rho_l h)^{0.5}$$

A probe may be inserted through a suitable hole to measure air velocity in a duct, but whatever instrument is used, it is important that the size of the probe should be small relative to the duct, so as not to reduce the flow significantly. Ideally there should be straight duct before and after the pitot tube of 10 diameters in length. As the air flow is not uniform across the duct, for best results a traverse should be made.

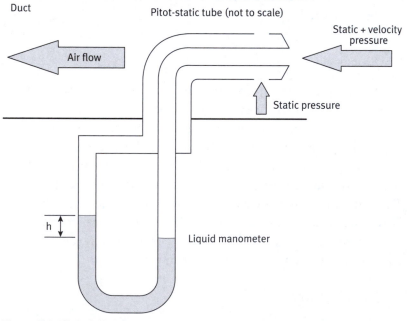

Figure 5.1 Pitot-static tube

Ventilation or infiltration rates

Air infiltration is the adventitious entry of outside air into a building, and may constitute a heat loss or a heat gain. It is therefore important to know the infiltration rate, which can be measured in two ways, using tracer gases or blower door tests.

Tracer gas method

One commonly used technique relies on the introduction into the room of a foreign gas which can be easily detected and whose concentration can readily be measured. Commonly used gases include CO_2 and N_2O, which are readily available in cylinders.

The valve is opened on the cylinder to allow some of the gas to be injected into the room, and when sufficient gas has been deemed to enter the room, the valve is closed. A fan may be used to distribute the gas uniformly. After injection has ceased, the concentration of the trace gas will begin to fall as fresh air enters from outside and air containing the trace gas exits. The rate at which the concentration falls is dependent on the rate at which the air in the room is changing, or in other words, the ventilation rate.

The rate of change of concentration of the gas depends on the amount of gas leaving the space, the amount entering, and the amount generated in the space.

$$V.dC/dt = G + Q(C_e - C)$$

V = volume of room
G = generation rate of gas in room (m^3/s)
C_e = external or background concentration of gas (kg/m^3)
C = concentration of gas in room at time t (kg/m^3)
Q = Quantity of outside air entering (m^3/s).

When injection of gas into the space has stopped, G = 0. (Note that if we use CO_2, G may not be zero as some will be exhaled by the person carrying out the test if they remain in the room.)

$$VdC/dt = Q(C_e - C)$$

If the external gas concentration is zero, Ce = 0.
(Note that C_e will not be zero if CO_2 is used.)

$$VdC/dt = -QC$$

Separating the variables:

$$dC/C = -(Q/V)\ dt$$
$$C_t = C_0\ e^{-(Q/V)t}$$

Taking logs:

$$\ln C_t - \ln C_o = -(Q/V)t$$

where C_o is the concentration at the beginning of the decay and
C_t is the concentration at time t.

$$Q/V = N = \text{Number of air changes per unit time}$$

Plotting ln C against t (in hours) gives a reasonable approximation to a straight line of slope –Q/V, which is the air change rate
in air changes per hour.

Blower door test

To assess the infiltration and leak characteristics of a building, a
blower door test can be carried out (see Figure 5.2). It is not
necessary to test all buildings, but where a leakage rate of $10m^3/hr/m^2$ or lower is claimed, then a test must be carried out to
provide confirmation. Full details are given in CIBSE Technical
Manual 23. The equipment consists of a false door which is
positioned in an outer doorway of the building, and which con-
tains a fan and flow measuring apparatus. The pressures inside
and outside the building are measured simultaneously. All the
outer doors and windows are closed but internal doors should
be left open. When the fan blows air into the room or building,
the air pressure inside will increase. Air will be forced out of the
building through small cracks at the edges of doors, windows,
floor–wall joints and so on. The greater the total air leakage

through these cracks, for a given flow rate, the lower the excess pressure will be. Thus, there is a relationship between the flow rate, the pressure build up and the leakage characteristics of the building.

A variable speed fan allows the flow rate and pressure to be measured over a range of flow rates. Pressure differentials of up to 50 Pa are normally used in order to obtain reasonable accuracy and flow rates. As these are far higher than encountered normally through wind pressure on the building, the results must be extrapolated to more reasonable pressure differences, or the raw values used directly for comparison with other buildings at the same pressure difference. A very approximate assessment of the annual infiltration rate can be obtained by dividing the leakage rate at 50 Pa by 20. The relationship between pressure and flow is of the form:

$$Q = k(\Delta P)^n$$

$$\text{Log}_{10} Q = \log_{10} k + n \log_{10} \Delta P$$

where

Q = flow rate (m^3/s)
ΔP = pressure difference (Pa)
k = constant
n = constant

Plotting $\log_{10}Q$ against $\log_{10}\Delta P$ should give a reasonable approximation to a straight line whose intercept is $\log_{10}k$ and slope n. Thus, values for k and n can be obtained and inserted in the

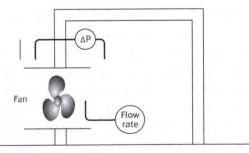

Figure 5.2 Blower door test

Average temperature of water from boiler to
    system   82 °C
Average return temperature to boiler   75 °C
Average water flow rate in heating system   1.6 kg/s
Average outdoor air temperature   7 °C
Average Indoor air temperature   23 °C
Gas consumption over the measuring period
    (6 hours)   35.5 m³
Specific heat of water   4190 J/kgK
Calorific value of gas   44 MJ/m³
Cost of gas   4.5 p/kWh

The efficiency of the boiler is equivalent to the heat output/heat input.

Heat output = $mC\Delta T$
= 1.6 × 4190 × (82-75) = 46,928 W

The heat input is based on the gas input multiplied by the calorific value.

Heat input over 6 hours = 35.5 × 44 MJ = 1562 MJ.

The input and output should be converted to the same units; in this case W are chosen.

The rate of heat input is 1562 MJ in 6 hours, so this must be converted to J/s.

1562 MJ/6 hours = 260.3 MJ/hour
=260.3 × 10⁶ J/hour
=260.3 × 10⁶/3600 J/s
= 72314 W.

Efficiency of boiler = 46,928/72,314 = 0.649 = 65%.

As the average outdoor and indoor temperatures are also known, it is possible to estimate the heat loss coefficient of the building.

Average heat output of the heating system (above) = 46928 W.
Average indoor-outdoor temperature difference = 23 - 7 = 16 K.
Total heat loss coefficient = heat loss/temperature difference =
    46928/16 = 2933 W/K.

Knowing this, it is possible to use the degree day method to estimate annual energy consumption, as shown in the Appendix. The method is approximate and does not take into account thermal storage in the building fabric, and any variation in internal heat gains, but in the absence of other methods it can yield useful information.

U-value measurement

U-values will normally be read or deduced from the building documentation, if available, but it may be desirable to measure them in some cases – where there is uncertainty as to whether insulation has been added to a building, for example, The thermal transmittance of a building element (U-value) is defined as the 'average heat flow rate per area in the steady state divided by the temperature difference between the surroundings on each side of a system' (ISO 7345). The units are $Wm^{-2}K^{-1}$. Accurate methods for measuring U-value use a guarded hot-box technique and are described in BS EN ISO 8990, but an approximate value can be obtained as shown in Figure 5.4 by logging the temperatures either side of the wall and the heat flow through it. The apparatus required includes a data logger, temperature sensors such as thermocouples or thermistors, and heat flow sensors. A heat flow sensor comprises a thin disc of material with a constant thermal resistance which has embedded in it a series of thermocouples to measure the temperature difference through the disc. A millivolt output is produced which is proportional to the heat flux. The heat flux and temperature difference should be logged at intervals of 10–15 minutes over a few days to minimize the effects of thermal storage. The average temperature and heat flow values over the period can be used to calculate the U-value, which is given by:

$$Q/A = U\Delta T$$
$$U = Q/(A\Delta T)$$

where

Q/A = the heat flux measured by the sensors (W/m^2)
ΔT is the temperature difference (K).

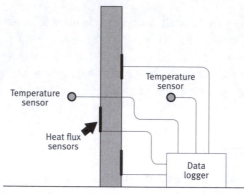

Figure 5.4 Measuring the U-value of a wall: the temperature and heat flux sensors

There are a number of potential sources of error in such measurements:

- Heat flow may not be one-dimensional, particularly near corners.
- Thermal storage effects in the wall may lead to erroneous readings.
- The section of wall chosen for the measurement may not be representative of the whole: for example insulation in the wall may be damaged or may have settled out.

They can be reduced by:

- positioning the heat flux sensors well away from potential disturbances.
- monitoring over long periods to reduce storage effects.
- making measurements in several places.
- using several heat flux sensors.

Data loggers

A data logger is a piece of electronic equipment with a series of channels which can be configured to accept inputs such voltage, current or pulses, and digital inputs, has a memory for storing the data collected, and facilities to download the data to a computer or printer. Once downloaded, the data can be processed using a spreadsheet or proprietary software.

A large range of models is currently available, and while some are designed for specific applications such as collecting electricity consumption data from a small number of input channels, others are more versatile and can accept various forms of data into as many as 60 input channels. Some specialized temperature and humidity sensors have integral loggers and are small enough to be unobtrusive when placed in a room. The frequency of monitoring may normally be varied in a range from once per second to once every 24 hours, depending on the variable being measured, and logging may often be carried out for months at a time if the memory is sufficiently large. Some loggers require manual downloading, while others may be downloaded remotely. Equipment may be purchased or hired, or a firm of energy management consultants may be appointed to carry out the logging and subsequent data analysis.

Errors in measurement

Care is needed, particularly in temperature measurements, to ensure that the sensor is measuring the quantity desired. A sensor intended to measure the ambient air temperature should not be exposed to direct solar radiation, as it will absorb energy and measure not the air temperature, but a combination of air and radiant temperatures, which could be more than 10 °C higher.

To measure air temperature, direct radiation should be excluded by placing the sensor inside a radiation shield, which can be as simple as a polished metal cylinder surrounding the sensor, large enough to allow free passage of air around the sensor. Alternatively the sensor should be mounted in a location where it is always in the shade.

Similarly, errors can be introduced into readings as a result of the response time of the sensor or instrument, but in general, energy-related variables in buildings change slowly compared with the speed of response of modern instruments. Temperature changes occur at a rate of fractions of a degree a minute, while sensors take only a few milliseconds to respond.

CHAPTER 6
ORGANIZATION AND IMPLEMENTATION

Some organizations may simply commission an energy audit, act on some of its findings, and forget about energy management for the next few years. While this approach may achieve some savings, for lasting and increasing savings a strategic approach is required. In today's energy markets the long-term view is essential.

For long-term effectiveness, energy management should be thought of as a continuing process, not as a single activity carried out once and forgotten. Making this happen requires commitment from all levels in an organization. In this respect it is important to develop an action plan, which helps to establish priorities for improving how energy is managed in the operation. Central to the action plan is a need to assess the organization's position regarding energy management, and an assessment of current operating practices is essential. The level of commitment to control of energy can be assessed and appropriate action taken in areas where there are deficiencies.

The organization of energy management takes various forms, depending on the size of the enterprise, the amount spent on energy, and the corporate structure. Many firms employ an in-house energy manager, who may be a specially appointed person whose sole responsibility is energy, or a member of staff who has other main duties. Alternatively, outside consultants may be employed. Energy management companies offer a range of services, including carrying out energy audits, running utility services such as combined heat and power (CHP) plants, financing investments in energy plant, or providing heat and electrical power. Forms of contract include fixed-fee arrangements and those where the consultant takes as their fee an agreed proportion of any savings, often on a no-savings, no-fee basis. Since many energy-saving measures tend to be long-term in nature with a number of years to payback, the latter tend to be long-term contracts of from five to fifteen years.

Whether the energy manager is based in-house or is an outside consultant, they need to:

- develop an energy audit system
- agree tough but realistic targets for energy savings
- provide technical advice
- monitor developments in energy conservation
- advise on government funding such as grants
- keep abreast of political, legislative and regulatory measures affecting energy use and costs.

Besides technical and commercial knowledge of buildings and energy supply, an energy manager also needs good interpersonal skills in order to be able to motivate management and other staff to invest in and implement energy-saving measures.

Functions of the energy manager

- Assess the level of awareness and commitment from management.
- Get commitment to energy management from the top of the organization.
- Identify the corporate structure and management style.
- Assess the level of awareness within the company generally, and raise it if necessary.
- Motivate others in the organization to help improve energy performance.
- Devise an energy policy.
- Assess current performance levels.
- Set up short, medium and long-term objectives and develop procedures to accomplish these objectives.
- Set up a system to measure and document energy use, and forecast future consumption.
- Agree targets for improvement, along with the associated budgets.
- Set up a programme to review objectives and achievements at regular intervals.

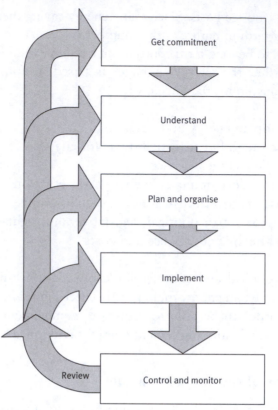

Figure 6.1 Organizing energy management

Assessing the level of awareness and commitment from management

It is important to know where the organization stands from the outset, and this can be ascertained by asking a number of key questions:

- Is there an energy policy?
- How aware of energy use are the workers and managers?
- Is there a budget for energy improvements, and how are they financed?
- Are there accurate records of energy use?
- Are there any plans for improving efficiency?
- What are the opportunities for intervention to improve efficiency?
- What are the risks?

Getting commitment

Most organizations with core activities such as financial services, manufacturing, software design or local authority services tend to regard energy as an unwelcome but unavoidable cost to the business, and have no particular interest in it. However, price rises in recent years have highlighted the true costs of energy, and they can no longer be ignored. In difficult economic times obtaining commitment to significant investment in energy efficiency is never going to be easy, and a careful and detailed economic analysis of the costs and benefits of the options on offer is the best way to convince management to invest. If a company has taken on a full-time energy manager then at least some commitment has been demonstrated, but managers might feel that paying the energy manager's salary is sufficient, and not understand that further investment will be required to generate significant long-term savings.

The following benefits should be stressed:

- Significant savings could be made with little or no capital expenditure.
- Good energy performance creates an image of the company as environmentally conscious.
- Staff may see a chance to receive or share some of the benefits and this will improve morale.
- Furthermore, every £1 saved is worth a full pound, while every £1 in increased sales produces only a fraction of that in profit.

The barriers that need to be overcome include the following:

- Energy is seen as a side issue, not part of the main business of the company, and diverts people from their main job.
- Lack of awareness.
- Energy consciousness is seen as a management cut-back and penny-pinching.

A further obstacle is that only 10 per cent of office buildings are owner-occupied, and this acts as a serious deterrent to investment in energy efficiency. Landlords may be unwilling to invest in measures which reduce their tenant's energy bills, while tenants have no incentive to invest in measures that might only pay off after they have left.

Money to invest in energy-saving measures is not usually handed over without careful scrutiny of the costs and benefits, using some of the techniques discussed later in this chapter. Investment in energy efficiency is often given low priority, as it is usually identified with savings rather than investment. There will normally be a number of projects competing for limited funds, and the organization will quite naturally wish to make the best use of its money. Financial controllers will be looking to obtain the optimum benefits from each investment with the minimum risk, and it will be essential for the energy manager to understand how the enterprise allocates money between capital and revenue budgets. Also, it may be difficult to identify an appropriate budget under which energy saving schemes could come. Capital plant, development or maintenance budgets may be suitable but, again quite naturally, it is often the case that the budget holders wish to protect their budgets for other uses and are unwilling to have their territory encroached on.

Opportunities

While there may be a small ongoing energy budget, other events may present opportunities to invest in energy-saving measures. These include:

- Single actions identified at particular times, e.g. refurbishment of a building, particularly when it involves a change of use; this may include opportunities to install CHP, upgrade the insulation of the building, install double glazing, or use natural ventilation in place of air conditioning.
- Programmes carried through an organization, such as a hotel chain upgrading the lighting throughout all its hotels. It may also be possible to hang other measures on the general programme, such as installing sub-metering or better controls.

Risks

There is usually some element of risk involved, but for most measures this will not be a technical risk (such as, will the insulation work?) but a financial risk – the extent to which the project is exposed to variations in factors which affect cash flows such as energy tariffs and interest rates. While energy costs follow a generally upward trend, there can be significant short-term variations, and committing to purchase large quantities at a particular time can sometimes lead to unwelcome losses.

Identify the corporate structure and management style

The way in which the organization operates will dictate the approach that has to be made to motivating people and obtaining appropriate budgets. The structure and management style may range from very hierarchical to strictly egalitarian. It is also important to identify those responsible for energy-consuming plant. Where this includes jointly used equipment such as photocopiers, it can be difficult to track down the responsible individual; it may be the maintenance/engineering department, the building manager, or a facilities manager. The implications are not trivial: a photocopier left on all night will use enough power to produce about 1500 copies. Consider the annual cost of this, and the cost for someone to switch it off before the office is closed.

Assessing and raising staff awareness

Raising general staff awareness of energy issues is an important element of obtaining commitment. Energy is in the news a great deal these days, but staff still need information on how energy use affects them in their organization and their particular job. A general 'save energy' approach does not necessarily help staff identify the areas where they can best help. A start can be made using a home energy assessment survey, to identify how aware staff are of energy savings that can be made in their own homes. Many of the lessons learned can be applied to the workplace.

Methods of raising awareness can take a number of forms, including presentations, training sessions, use of company newsletters, email and websites, suggestions schemes, poster campaigns, and 'save it' stickers. The ultimate aim is to integrate energy efficiency into everyday working practices and make energy-saving activities such as switching off lights part of the everyday routine.

Motivating others

Raising awareness is part of motivation, but motivation also includes other aspects such as personal, departmental or team reward schemes for improved performance, linked to monitoring and targeting activities. Motivating management may require demonstrations of potential energy savings using detailed costed examples, as budgets will need to be agreed. People will generally not change to work in an energy-efficient way for its own sake – they need to see that there is some benefit to themselves for doing so. Such benefits could include greater profitability and competitiveness for the company as a result of lowering costs, and better job security. Benefits to individual departments, such as increased budgets or a chance to utilize a share of the energy savings, may also act as an incentive. Schemes such as 'energy saver of the month' and suggestions schemes have also been shown to be effective. An important point here, as with raising awareness, is that a 'one-off ' approach tends to fall off in effectiveness after some time, and new approaches or initiatives need to be introduced from time to time in order to maintain the momentum. Communicating the results of any initiatives back to those involved is also crucial in maintaining interest. Articles in a company newsletter or on the website can highlight improved performances of groups and individuals.

Devising an energy policy

A company's energy policy needs to be agreed with senior management since its implementation will involve important investment decisions. The objectives of the energy policy will depend to some extent on the current attitudes towards energy, and the existing energy performance within the organization.

The following steps may be taken.

Assess current level of performance

It is essential to know how well the organization is succeeding at present in the management of energy:

- Are space temperatures appropriate, too hot or too cold?
- Are lights left on unnecessarily?
- How well are the building fabric and services plant maintained?
- Is electrical equipment routinely left on overnight?

The answers to these questions will point the way towards short-term improvements that will help to inspire confidence in the energy management process.

Setting up short, medium and long-term objectives

It is important to set up specific objectives rather than vague commitments to 'cut energy' or 'be more efficient'. These objectives will vary depending on the time scale being considered. A simple low-cost change that gives a quick result will be very effective in developing confidence in the energy management process, and it may be possible to use the savings to finance more long-term measures.

Examples of short medium and long-term objectives

Short term

- Mend faulty items of energy plant.
- Ensure items of equipment are switched off when not in use or at the end of the working day. Simple time switches to turn items off at the end of office hours will pay for themselves in weeks.
- Ensure rooms are not heated or cooled excessively.
- Sticker campaign for light switches.
- Report back on improved performance.

Medium term

- Set a target of 10 per cent energy reduction in a year.
- Instigate a better maintenance regime.
- Report on costs and benefits of sub-metering.
- Appoint local 'energy champions.'

Long term

- Set up a five-year energy reduction target of say 20 per cent.
- Set up monitoring and targeting software for cumulative sum (CUSUM) analysis and similar activities.
- Report on feasibility of replacing boilers with CHP.
- Train energy champions.

Setting up a system to measure and document energy use

Energy consumption information must be authoritative if it is to be used to inform the investment policy: that is, it must be obtained from reliable sources, and accurately documented. Such sources include detailed utility bills, half-hour metering information, and logged data from the building management system (BMS). Future consumption can be forecast using such techniques as the CUSUM analysis described later in this chapter.

Agreeing targets for improvement, along with the associated budgets

The measures required to meet the energy reduction targets should be properly costed, and the benefits presented realistically and accurately, otherwise management is unlikely to agree to release funds. Targets which are imposed, not agreed, will have little effect since no one will feel committed to honour them.

Setting up a programme to review objectives and achievements at regular intervals

The appropriate time scale for review should be selected. Too short a time scale will not allow any savings to be made or demonstrated, while if too long a period is chosen, people will lose interest. If new plant is installed, such as a condensing boiler, it will be some months before the benefits are evident in the energy bills. With other measures it is likely to take a matter of years for the benefits to materialize, therefore it is pointless to review the objectives before the measures have had time to take effect.

The plan should include a regular commitment to review progress, as needs may change over time, for example due to change in use of parts of the building. It is important that regular reviews are fed back into the entire energy management process in order to take into account changes in energy supply, legislation, changes within the company, and finance. A large amount of free literature on this subject is available from the Carbon Trust (www.carbontrust.co.uk).

Working to an agreed standard is a useful way of achieving and demonstrating commitment. BS16001:2009 shows in detail the requirements to be met in order to achieve the British Standard for Energy Management, and enlarges on the outline of organized energy management shown here.

Monitoring and targeting

Monitoring and targeting (M&T) form essential elements in understanding and controlling energy use, and also provide information to feed back to management on the performance of the building. The purpose of M&T is to relate energy consumption to some variable such as the weather or production output, in order to understand better how energy is used and help identify avoidable waste.

Data collection is an essential element of M&T, and may be carried out automatically or manually, depending on the kind of data being collected, the technology available, and the budget. Analysis of the data will suggest clear lines of investigation for producing savings, which can easily be quantified.

It is important to set targets that are realistically achievable, but not so low as to be meaningless. An overall reduction of 5 per cent in energy costs over five years is unlikely to justify the time and money put into it, while a target of 30 per cent is extremely tough if no special measures are taken, but may be realistic if a budget for significant new plant such as CHP is available. When an organization is just beginning an energy management programme, targets of 5–10 per cent may often be achievable at low or zero cost by simple common-sense measures such as switching off unwanted items of plant, but these are one-off items that are unlikely to provide further opportunities.

Targets should be discussed and agreed between the relevant parties, and based on sensible calculations of performance. The smaller the unit being targeted the better, as greater control can be exercised, and frequent reviewing of performance enables problems to be identified quickly. Examples of such problems include faulty timers allowing heating to remain on 24/7 even over Christmas holidays, badly calibrated temperature sensors or broken control valves.

A key element of M&T is accurate forecasting of the expected energy consumption; exceptions can then be reported and acted upon. A simple way of reporting exceptions is to use an overspend league table, which shows the overspend on specific items of energy consumption over a fixed period, in descending order of cost. High overspend items can be discussed with those responsible, and further analysis carried out, or if the reason for the overspend is obvious, any necessary remedial action taken. A week is often a suitable reporting interval, but daily or monthly reporting may be used, depending on the circumstances.

Calculations of expected consumption are based either on precedent (direct comparison with previous periods) or activity: in other words related to the driving factors, such as weather or production quantity. Precedent-based targeting is usually based on monthly figures and year-on-year comparisons, and suffers from the weakness that significant changes in the weather may occur from one year to the next, which are not taken into account in this method. It also does not take into account changes in working practices such as opening hours and weekend working. This makes the determination of 'exceptions' rather difficult.

Activity-based targeting enables changes in the weather to be allowed for, and as we now have relatively easy access to weather data from a large number of locations, it can be used to provide more incisive analysis of the performance of a building. Heating and cooling are among the greatest causes of energy use in buildings, and are related directly to the weather. Plotting a graph of heating energy use and degree days as shown in Chapter 3 is one of the simplest but most instructive ways of analysing performance. M&T software can be configured to produce such plots automatically. For plots of energy use against degree days the relationships are linear, of the general form $y = mx + c$, where m is the slope and c is the intercept on the y axis. Similar plots can be drawn for process energy, such as energy use against throughput for an oven or drying plant, but they are not invariably linear. Process energy plots may in fact reveal that consumption is unrelated to throughput, but this knowledge in itself is useful as it may point to a better form of operation or control in order to save energy. For batch processes, plant is often left on between batches, which could lead to large amounts of energy being wasted.

When making such plots the user normally has the option to set the intercept at zero. For heating degree days this will not usually be the case, and the intercept will depend on a number of factors, including the base temperature used for the degree day calculation. For processes there may be a fixed amount of energy use plus a throughput-related element (c and m respectively in the $y = mx + c$ plot).

It is also essential to be able to measure the extent to which the target has been approached. This is done by analysing utility bills, but may also be done by monitoring, using some of the instruments and techniques described in Chapter 5. Smart meters which provide this kind of data for the customer and also to the energy supplier are now becoming more common, and will be universal for domestic properties in the United Kingdom by 2020. They may also be linked in with automatic meter reading (AMR) systems which have been trialled in a number of places. An essential feature of both these is that half-hour meter readings can be made, giving both the supplier and the user much finer-grained data than was previously available. For large users half-hour meter reading is essential since the cost per unit varies with time of delivery; such a detailed knowledge of the consumption pattern may give the user the opportunity to shift loads to cheaper times of day. It may also point the way to avoiding or minimizing maximum demand charges by shifting some loads to times of lower demand. In addition, BMS may be configured to issue alarms when certain limits are being approached.

A conventional meter costs about £500 including installation, and if a payback of two years is expected, then each meter needs to save £250 per year. A minimum of 5 per cent savings should be aimed for, so that the utility cost through each meter installed should be £5000 to make it worthwhile.

Analysis of the data using CUSUM or other methods may be done using proprietary software (some of which is available as an add-on to the BMS) or by the energy manager generating a bespoke spreadsheet.

CUSUM analysis

CUSUM stands for the cumulative sum of deviation: in other words, deviation from the expected consumption. CUSUM analysis is a tool of M&T, and in its simplest form the cumulative sum of energy consumption is plotted against time. The data required can be obtained from energy bills, from BMS data, or from a monitoring exercise. The slope of the curve will change from time to time, depending on the conditions. For example, gas consumption may cover heating, cooking, and domestic hot water, and in summer there will be no heating, so the consumption will be much reduced and the slope will be lower (see Figure 6.2).

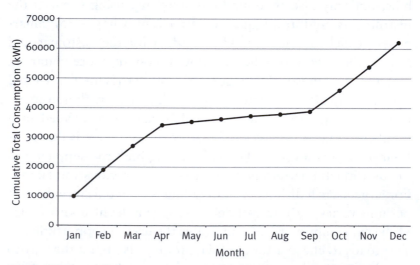

Figure 6.2 Cumulative energy consumption (total of space and water heating and cooking)

Table 6.1 Monthly heating gas consumption and degree days

Month	Previous year heating gas consumption kWh	Previous year degree days DD	This year heating gas consumption	This year degree days
Jan	9100	598	9069	576
Feb	8765	488	8054	453
Mar	7250	354	7073	324
Apr	6200	269	6329	272
Oct	6243	333	6154	327
Nov	6989	442	7078	434
Dec	7600	528	7509	524
total	52147	3012	51266	2910

The slope effectively gives the rate of energy use. Since there is no space heating in summer the slope reduces considerably during those months, and is approximately constant over that period, indicating that the cooking energy is roughly constant. Ideally data from more than one year should be available, when a straight comparison from one year to another can be carried out, allowing the seasonal changes in consumption to be observed (see Table 6.1).

The results are best shown in histogram form (as in Figure 6.3). Here, only the raw heating energy data is used and differences in the weather over the same period are not taken into consideration. A simple way to take account of the weather is to calculate the energy consumption per degree day (see Chapter 3), and if kWh/degree day are used, then a direct and meaningful comparison can be made (see Table 6.2 and Figure 6.4).

Table 6.2 Energy consumption in kWh/DD for current and previous year, deviation and CUSUM

	kWh/DD			
	This year	Last year	Deviation	CUSUM
Jan	15.74479	15.21739	0.5274	0.5274
Feb	17.77925	17.96107	-0.18182	0.345584
Mar	21.83025	20.48023	1.350021	1.695605
Apr	23.26838	23.04833	0.220055	1.91566
Oct	18.81957	18.74775	0.071824	1.987484
Nov	16.30876	15.81222	0.496539	2.484023
Dec	14.33015	14.39394	-0.06379	2.420236

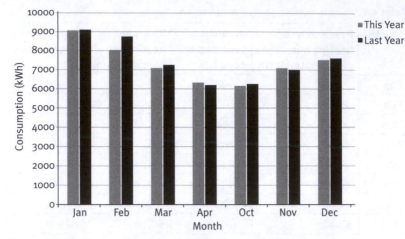

Figure 6.3 Heating energy consumption for two years compared

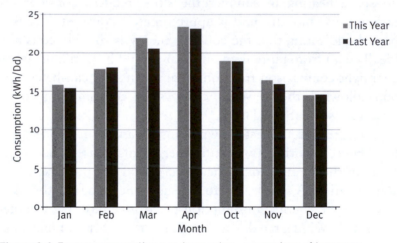

Figure 6.4 Energy consumption per degree day – comparison of two years

CUSUM analysis is relatively simple to perform, as it can be carried out using spreadsheets or standard M&T software. It can be used in three main ways.

- to set realistic targets for energy consumption
- to assist in the diagnosis of persistently excessive consumption
- to quantify cumulative savings.

The analysis assumes that if the building is operated correctly, energy consumption will be predictable. Calculation is based on energy consumption over a specified measuring interval: monthly, weekly, daily, or even hourly intervals can be used depending on the requirements. Whatever interval is selected, at the end of each period the actual quantity of energy used must be measured and an estimate made of the amount of energy that should have been used: in other words, the expected consumption. The deviance for the period is obtained by subtracting the expected from the actual consumption. Adding the deviance for the last period to the running total of deviance gives the CUSUM, or cumulative sum of deviation.

The example on the next page shows the operation of CUSUM analysis in the context of heating a building. The actual consumption data can be obtained from meter readings, but it is more difficult to calculate the expected consumption. In the case of manufacturing industry a useful measure of expected consumption is the amount of product produced per unit of energy consumed: for instance for a steel works, tonnes per GJ, or for vehicle use, miles or tonne miles per gallon. If output increases we would expect energy consumption to increase in a predictable way. For the building applications being considered here, output is rather too abstract a concept to use, since many of the kinds of buildings we are considering – houses, offices, leisure centres and so on – do not necessarily have an 'output' which is easily measurable. In any case, the 'output' may not be something which can meaningfully be compared with the energy consumption. For example the 'output' of a public swimming pool might be the number of customers per day, but the swimming pool would still have to be heated irrespective of whether there were 3 or 300 customers using the pool in a day. The amount of energy needed is less dependent on the number of customers than on other factors, such as the weather. In this instance where we are concentrating on heating energy use it is clearly the weather, as measured by degree days, that is the main driving force of energy consumption. (For details see the Appendix.) For summer cooling a corresponding measure, the cooling degree day, can be used. In a well-regulated building the heating energy consumption per degree day should be roughly constant.

Figure 6.5 shows the data for one year from Table 6.2 redrawn in this way. Since the heating energy consumption is zero over the non-heating season, points from this period are

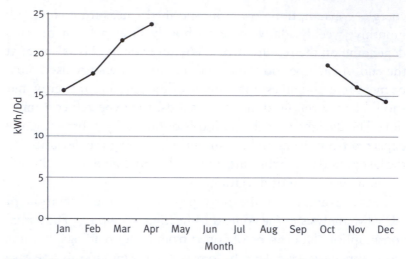

Figure 6.5 Heating energy consumption per degree day, plotted by month

omitted. (Energy consumption divided by zero degree days would give infinity.)

This form of plot already yields additional information. It may be observed that the energy consumption per degree day increases from winter into spring, and decreases from autumn to winter. This may occur for a number of reasons. One is that in winter the load factor will be greater and the boilers will be running more efficiently; another possibility is that in the coldest weather the occupants are more careful about closing doors and windows.

In order to pursue the CUSUM analysis the expected energy consumption must be calculated. The usual procedure is to take the average consumption per degree day for the previous year, calculated here as 17.3 kWh/DD. Using this figure the expected energy consumption for the corresponding period in the following year and the deviation can be calculated. These values are shown in Table 6.2 and Figure 6.6.

The CUSUM analysis is shown in graphical form in Figure 6.4. It may be observed that for most months the deviation is increasing, demonstrating that consumption is worse than the previous year. During two months the deviation decreases, illustrating better performance than previously, but the overall trend is upwards away from zero. A sharp increase in deviation between February and March indicates that something serious

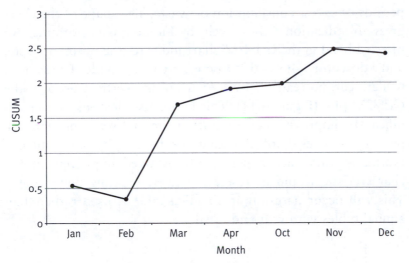

Figure 6.6 CUSUM analysis for the example given

has occurred – a broken sensor or valve, for example. During the summer months the deviation falls considerably, suggesting that the main problems lie with the space heating. From November to December the deviation falls, possibly following some remedial action to improve control. The cumulative deviation at the end of the year is significant, showing that there has been an overall deterioration in performance since the previous year. The reasons for this may be manifold, and detailed study of the CUSUM analysis is of help in finding them.

In a well-controlled building the 'expected' consumption for this year should be roughly the same as the previous year, after allowing for degree day differences. Although inevitably there will be slight positive and negative deviations, the expected consumption will follow a line roughly parallel with the x-axis.

If the CUSUM from the current performance characteristic is correctly derived and the building is well regulated, the CUSUM chart will be similar to that in Figure 6.7. There are slight variations from the norm from month to month, but on average the trace runs level. If the characteristic has been set too leniently, there will be a general downsloping tendency (Figure 6.8). On the other hand, if the characteristic is set too harshly there will be a persistent rise in the graph (Figure 6.9). Figure 6.10 shows a typical CUSUM chart for a building which has a correctly set target, but like most buildings, has occasional periods of excess consumption or waste. The shape of the chart gives an indication of how well the building is performing. An upward bend in the CUSUM chart indicates the onset of waste, and a downturn shows that savings are being made. The savings to date can be read off by measuring the vertical drop of the CUSUM plot (Figure 6.11). This figure also demonstrates a significantly improved performance from halfway through the period, as a result of changing the heating system controls. Normally after such a period of sustained improvement the characteristic would be reset, effectively creating new targets. This will occur naturally if a rolling table is used rather than annual tables with year-end breaks.

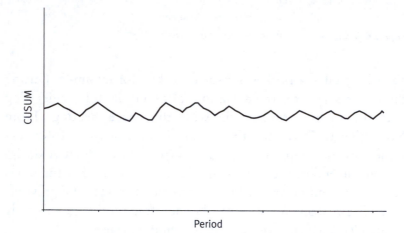

Figure 6.7 CUSUM of well-regulated building

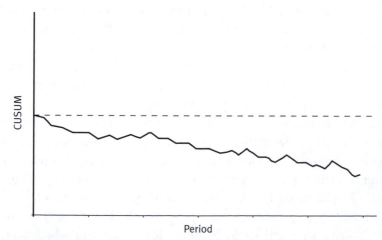

Figure 6.8 CUSUM of characteristic that is too lenient

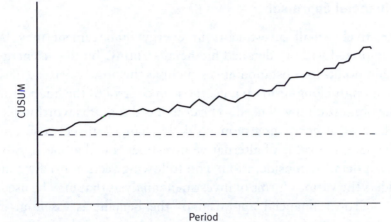

Figure 6.9 CUSUM where the characteristic is too harsh

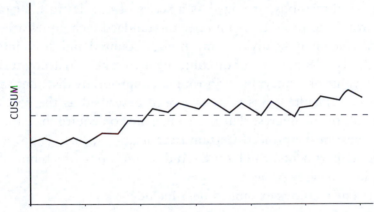

Figure 6.10 CUSUM with generally good performance but periods of waste

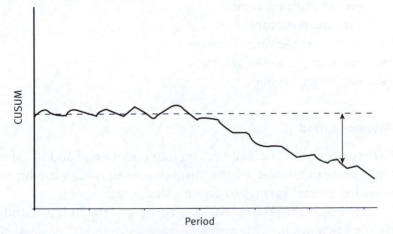

Figure 6.11 CUSUM showing cumulative savings (arrow

Financial appraisal

Relatively small investments in energy management may be authorized without detailed financial scrutiny, but if the energy audit points to substantial investment as the only way to achieve meaningful long-term savings, those in control of the finances in the organization will need to be convinced that it is worthwhile. The cost of borrowing money, likely changes in fuel cost, and the potential value of alternative investments will all need to be given detailed consideration. The following section briefly considers the various forms of investment analysis that may be used.

Proper financial appraisal of the benefits is essential in order to identify projects which make best use of the organization's money. Many large organizations will use a number of appraisal methods, arranged as a series of hurdles to filter out unpromising projects. A number of standard techniques are in use, which may be divided into non-discounted and discounted methods. The purpose of discounting is to take into account the time value of money, but choosing an appropriate discount rate can be difficult. It has variously been described as the cost of capital, or the interest that has to be paid to acquire the capital to invest in the project. Certain organizations, such as government departments, will have fixed discount rates which must be applied to every project.

The methods examined here include:

- payback period
- gross return on capital
- net return on capital
- gross average rate of return
- net average rate of return
- net present value.

Payback period

Payback period is the simplest of all to understand and to calculate. The capital cost of the project is simply divided by the expected annual savings to give a value in years.

The advantages are that it is easily understandable, and a simple calculation.

The disadvantages are that it does not take into account the timing of costs and benefits, likely residual value of assets at end of project life, or savings accruing after the payback term (see Table 6.3 and Figure 6.12).

Example of the payback method

Table 6.3 Simple payback method calculation for two energy-saving measures

Year	Measure A			Measure B		
	Cash out	Cash in	Cum net cash in	Cash out	Cash in	Cum net cash in
1	1000	0	-1000	1500	0	-1500
2	0	200	-800	0	300	-1200
3	0	200	-600	0	300	-900
4	0	200	-400	0	300	-600
5	0	200	-200	0	300	-300
6	0	200	0	0	300	0
7	0	200	200	0	300	300
8	0	200	400	0	300	600

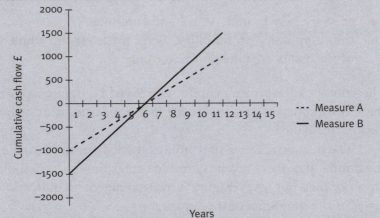

Figure 6.12 Cumulative non-discounted cash flow of two energy-saving investment measures with the same payback period

Measure A requires investment of £1000 and gives savings of £200 per year. Measure B requires £1500 investment, giving savings of £300 per year. While both measures have the same payback period, with the larger investment the cumulative savings after the payback-period are much higher, therefore on this simple basis it would be considered more worthwhile. The cost of paying back money for the investment over a number of years, possibly with varying interest rates, is not taken into account here.

Accounting rate of return

This is also known as the average annual rate of return on investment, and concentrates on profitability, taking no account of cash flow. It considers earnings over the entire life span of the project but does not take into account the timing of those earnings.

The two basic variants to this method are:

- average gross annual rate of return
- average net annual rate of return.

The average gross annual rate of return is defined as the average proceeds per year over the life of the assets expressed as a percentage of the original capital cost.

The average net annual rate of return is defined as the average proceeds per year, after allowing for depreciation over the life of the assets, expressed as a percentage of the average value of the capital employed.

The advantages of these methods are that:

- they are easy to understand and compute
- they emphasize profitability, as returns over the whole life of the assets are taken into account.

The concept of return on capital employed is often considered the most important yardstick used in the measurement of business performance.

The disadvantage is that no acknowledgement is made of the timing of costs or receipts, and irregularities in the cash flow are smoothed out by averaging. Once averaged, no indication is given of the time span of the return.

Examples of rate of return calculations

1. Gross return

Capital cost of project	£120,000
Total net cash inflow over 5 years	£170,000
Average net cash inflow per annum	£34,000
Average gross return (capital cost £120,000)	34,000/120,000
Average gross return	28.3%

2. Net return

Capital cost of project	£120,000
Residual value of assets at end (of 5 years)	£50,000
Total net cash inflow (over 5 years)	£170,000
Less depreciation	(£70,000)
Net return	£100,000
Average net cash inflow per annum (5 years)	£20,000
Average net return on average capital employed (£70,000)	20,000/70,000
Average net return	28.5%

Discounted cash flow (DCF)

DCF methods of investment appraisal involve discounting future outflow and inflows of cash back to present day values, thus establishing a common base for the comparison of investment alternatives. The difference between these and the methods already described is that they acknowledge the importance of timing, so that funds invested in the future have less impact than funds invested now, and that funds received early on in the lifetime of a project are worth more than funds received later. There are two basic DCF methods:

- net present value
- internal rate of return.

The present value of the funds invested is compared with the present value of the net cash flows expected to be generated over the life of the investment. An example is shown in Table 6.4.

The following information is required:

- initial cost of the project
- cost of supplying the capital required, that is, the minimum rate of return required from the investment
- the values and timings of future cash flows for the total expected life of the project
- reference table of discount factors.

The method of computation is as follows:

1. Select the appropriate discount factor.
2. Calculate the present day value of each year's cash inflow by multiplying the values of those inflows by the appropriate discount factors.
3. Sum the values calculated to give the total present day value of future cash benefits.
4. Calculate the present day values of future cash outflows and add the sum of these to the initial cost of the project.
5. If the total arrived at under 3, the total present day value of all future returns, exceeds the total arrived at in point 4, the present day value of the investment costs, the project should be accepted, and if not, it should not be pursued.

The method is particularly useful where several alternative projects are being considered, as the one showing the highest positive result is the best performer and is likely to be the one accepted.

Example of a discounted cash flow calculation

Three projects all have the same initial cost, and the same total cash flow in: the difference is in the timing of the cash flows. In A the greatest inflows are at the beginning, in B at the end, while in C the cash flow in is the same each year.

Project A

Table 6.4 Discounted cash flow example

Year	Net cash flow	Discount factor @10%	Present day value
0	(100,000)	1.0	(100,000)
1	50,000	0.909	45,455
2	45,000	0.8264	37,188
3	35,000	0.7513	26,295
4	30,000	0.6830	20,490
5	20,000	0.6209	12,418
Total cash flow	+80,000	Total NPV	+41,846

Project B

Year	Net cash flow	Discount factor @10%	Present day value
0	(100,000)	1.0	(100,000)
1	10,000	0.9019	9,091
2	15,000	0.8264	12,396
3	35,000	0.7513	26,295
4	50,000	0.6830	34,150
5	70,000	0.6209	43,463
Total cash flow	+80,000	Total NPV	+25,395

Project C

Year	Net cash flow	Discount factor @10%	Present day value
0	(100,000)	1.0	(100,000)
1	36,000	0.9019	32,728
2	36,000	0.8264	29,750
3	36,000	0.7513	27,047
4	36,000	0.6830	24,588
5	36,000	0.6209	22,352
Total cash flow	+80,000	Total NPV	+36,465

From the example shown in Table 6.4, it is seen that all projects provide the same net cash inflow, but Project A, where the greatest influx of cash is at the beginning, has the highest present day value, and on this basis would be the one selected.

Internal rate of return (IRR)

This is a variation of the net present value method except that it is used when the cost of supplying the capital is unknown or uncertain. It is particularly useful for indicating the most profitable of several alternative projects.

It requires the same data as the NPV method, except that the start point is an assumed 'break-even' total NPV: the method is then worked backwards, using trial and error, to find the discount rate which, when applied to the annual cash flows, produces the break-even result. The discount rate thus arrived at is the internal rate of return, and on this basis the project showing the highest rate is the most profitable one. These computed rates of return are then compared with the enterprise's existing rate of return on its present investments, or against the cost of providing funds, to assess whether or not the new project is worth undertaking.

Using the previous examples, it can be shown the internal rate of return for the three projects is:

Project A	27%
Project B	20%
Project C	22%

Summary of the NPV and IRR methods

- Both methods indicate whether a project is acceptable or not compared with the minimum acceptable rate of return or the expected finance cost of funds applied.
- Both methods indicate a preferential ranking of alternative projects, on the basis of cash flow (NPV method) or profitability (IRR method) respectively. The higher the rate of interest applied, the less valuable are cash inflows received later, and the less the impact of cash outflows incurred later.
- The NPV method assumes that the net cash inflow generated during the course of a project is reinvested at an interest rate no lower than that used as the discount factor. The IRR method assumes that the net cash inflow generated is reinvested at the IRR.
- The IRR method produces problems of computation where the cash flow pattern is irregular: that is, when cash outflows occur at future times in between the normal cash inflow occasions.
- Both methods involve an assessment of the cost to the enterprise of the capital it will use for investment. This in turn requires determination of the source of the funds to be used.

Another way of discounting is the annual equivalent cost (AEC). Whereas the NPV is the amount by which any future benefits at present value exceed the cost of the project, the AEC is the average amount by which the projects exceeds this in each year of the project's lifetime.

Sensitivity analysis

This is the process by which key design features are tested to determine what impact they may have on the project. The first step is to identify those components of the capital cost for which there is a margin of error, such as installation cost, or the variable cost of components sourced abroad. Sensitivity tests may also be applied to the benefits, such as errors in the estimate of post-implementation costs, particularly those caused by variations in fuel cost or the weather, or the lifetime of the measure.

Financing the investment

A range of investment routes is available, including investment from within and borrowing from banks. A further possibility is equipment supplier finance, in which a third party buys the equipment and takes on the responsibility of having it designed installed, operated and maintained.

CHAPTER 7
LEGISLATION AND GRANTS

Introduction

Successive UK governments have adopted three main strategies to reduce energy consumption and carbon emissions – encouragement, in the form of grants, low or interest-free loans and tax incentives; compulsion in the form of regulations; and taxes on fuels. Since these measures tend to be amended at regular intervals, it would be fruitless to go into great detail here, and only a brief outline is presented. Surveys have shown that the main incentive to reduce energy consumption is cost. However, this alone appears to be insufficient to deliver the carbon reductions required, and measures have been taken to force people to reduce their fossil fuel consumption.

Warm Front scheme

On the domestic side, the Warm Front scheme helps poorer households to claim benefits to help upgrade the heating and insulation in their homes. Grants are currently available of up to £3500, covering such items as loft insulation, cavity wall insulation, draught proofing and heating.

Feed-in tariffs

At present there are no direct grants from central UK government for the installation of solar photovoltaic (PV) panels, although some local authorities may have schemes in operation. Feed-in tariffs (FITs) are designed to promote the use of solar

PV and other renewable electricity sources, by paying the owner not only for power exported to the grid, but for the power they generate and use themselves. The system is administered by OFGEM. Those eligible must produce less than 5 MW per year and have systems installed (by an accredited installer) after July 2009. Systems installed earlier may be eligible for a reduced rate. The tariff is paid for all the electricity generated, and varies depending on the source and the scale of generation; it also includes (at a much lower rate) existing generators transferred from the older renewable obligation (RO) scheme. The tariffs are set for up to 25 years, and go up to a maximum of 43.3p/kWh for retrofitted solar PV up to 4 kW. The scheme includes PV, hydro, wind turbines, micro-combined heat and power (CHP) (less than 2kWe) and aerobic digestion. The tariff is also index-linked and free from income tax. An export tariff of 3.1 p/kWh above the generation tariff is paid for electricity exported to the grid. Smart meters will be needed to measure the amount generated, used and exported, and systems have to be registered with OFGEM.

Renewable Heat Incentive (RHI)

Although 47 per cent of greenhouse gas emissions in the United Kingdom are attributable to heating, only 1 per cent of heat energy comes from renewable sources. The government's plans require that by 2020 15 per cent of energy should come from renewables, and it has been estimated that they could supply 12 per cent of the heat demand. The RHI is designed to encourage the production and use of renewable heat sources by providing funds to owners for every kWh produced. The sources funded (at different rates) include biomass, ground source heat pumps, air source heat pumps and solar thermal. A premium will be paid for heat exported, but it must be demonstrated that the source is connected to a heat network – otherwise it would be possible to collect income by generating unwanted heat and wasting it – and certain quality standards must be met. Tariffs up to 18 p/kWh are available, depending on the type of renewable and size, and will last for up to 20 years.

Example

A 6 m² roof-mounted solar thermal panel in the north of England costs around £6000 and produces 2705 kWh/year. Under RHI the owners can claim 18 p/kWh = £2705 x 0.18/year = £486.90 per year for 20 years, i.e. a total of £9738.00, greater than the cost of the panel. This is in addition to the savings made by offsetting fossil fuel use.

These amount to (assuming gas used at 80 per cent efficiency and 4.5p/kWh):

£2705 x 0.045/0.8 = £152/year

= £3040 over 20 years, a total saving of £12,778.

Payback period (PBP) without RHI = 6000/152 = 39 years.
PBP with RHI = 6000/638 = 9.4 years.
Carbon dioxide (CO_2) savings 498 kg per year.

With RHI and a ground source heat pump (GSHP), the example in Case study 2 (see page 147) would cut the PBP of the GSHP from 11 to 5.2 years.

Building Regulations

Most developed countries either have or are producing building codes to limit heat transfer through individual building elements or overall energy consumption.

Part L of the Building Regulations in England and Wales (Section 6 in Scotland) lays down the required standards for U-values, air-tightness, boiler efficiency, heating and hot water systems and controls, metering, and light fittings. Calculations are required to demonstrate compliance. Many of the regulations only apply to new build, but increasingly stringent standards are being applied to refurbishments, extensions and conversions.

Some of the practical implications are as follows.

Replacement windows are required to conform to the new U value standards (1.8 – 2.2 W/m²K), implying the use of low-emissivity rather than plain glass. The thermal performance of an extension will normally have to meet that of new build.

New U-values for walls at the time of writing are 0.19 W/m²K.

Air leakage targets are also becoming stricter – designers will be required to demonstrate that the building meets a target of 10 m³/h/m² at an applied pressure of 50 Pa by using a blower door test.

The regulations also now make reference to the reduction of cooling energy to encourage designers to introduce such passive systems as night cooling and increased thermal mass for damping of heat gains.

Local authority requirements

Many local authorities have now set up planning regulations concerning new buildings in their area which are more stringent than the Building Regulations requirements. The borough of Merton led the way, insisting that 10 per cent of the energy to be used in the building should be generated from renewable sources; other authorities have now adopted this rule, which has become known as the 'Merton rule'. It has been argued by some that a building would be more carbon-efficient if instead the loads were reduced significantly, rather than generating unnecessary power. Indeed, these local authority requirements have largely been superseded by other measures including the Code for Sustainable Homes and feed in tariffs, which provide much better incentives.

European Directive on Energy in Buildings

This came into force in the United Kingdom in January 2006, affects domestic and non-domestic buildings, and was introduced to help meet Kyoto commitments. Amendments were made in 2010.

The main manifestation in our buildings has been the display of energy performance certificates (EPCs). These have to be produced by a certified assessor. The Building Research Establishment (BRE) and others offer courses through which engineers and others can become certified. The calculation method used is the simplified building energy model (SBEM).

Main requirements

- Minimum energy performance for all new buildings (calculated by a prescribed method).
- Minimum energy performance for large existing buildings subject to major renovation.
- Energy certification for all buildings.
- Regular mandatory inspection of boilers and air-conditioning systems in buildings.

Public buildings are required to display the energy certificate, and it is perhaps too early to say whether this has had the intended effect of encouraging greater energy efficiency.

Climate Change Levy

The Climate Change Levy (CCL) came into force on April 1 2001, following the recommendations made in Lord Marshall's report *Economic Instruments and the Best Use of Energy* (October 1998) and two years of consultation with industry. It is a tax on energy use in industry, commerce, agriculture and the public sector. All UK businesses and public sector organizations pay the levy via their energy bills, but fuel oils do not attract the levy as they are already subject to hydrocarbon oil duty. The effect is to increase energy bills by 10 per cent or more. There is an additional £150 million of government assistance to business for energy efficiency measures.

CHP may be exempt or attract less CCL compared with conventional generators, depending on a number of factors including the scale of generation, the quality index (QI), and power efficiency (PE) (see Chapter 4).

Energy from renewable sources is exempt from the levy, and energy-intensive users who sign a climate change agreement (CCA) with the Department for Energy and Climate Change are eligible for a reduction of up to 80 per cent.

Small businesses paying VAT at a reduced rate are automatically exempt from the levy. One of the original intentions was to aid manufacturing industry by raising the price of energy relative to labour, and it was therefore expected to have a favourable impact on employment. The revenues raised are recycled to business through a 0.3 per cent reduction in employer's National Insurance contributions. Overall, the levy is intended to be revenue neutral: in other words, the amount the government gains in levy is paid in out in a reduction in the NI contributions. There have, however, been criticisms from manufacturing industry that some sectors lose out considerably. Businesses that use large amounts of energy but have few employees (such as Scotch whisky distilling) are net losers, while employers that use relatively small amounts of energy but have a large number of employees, such as the Royal Mail letter services, are large gainers.

Tax revenues are recycled into support for energy efficiency, and heavy industrial energy users can obtain partial rebates by entering into energy efficiency agreements with the government. The CCL in the United Kingdom is administered by HM Revenue and Customs, to which application should be made for registration and exemption. It was expected to deliver a saving of approximately 5 million tonnes of carbon a year by 2010, and there is some evidence that the effect of the levy on fuel price, particularly electricity, has encouraged greater efficiency.

Emissions trading schemes

The UK Emissions Trading Scheme (ETS) was voluntary, and ran as a pilot scheme for the European Union's ETS (EUETS), which covers over 10,000 installations, which together are responsible for almost half the EU emissions of CO_2. The rules apply mainly to those in the industrial sector, such as iron and steel, cement, glass, ceramics and paper manufacture. Large emitters must monitor and report their emissions, and must return emission allowances equivalent to their emissions, on a yearly basis. If their emissions are higher they may purchase more credits, or sell them if emissions are lower. Thus, credits may be traded on an exchange. It was a policy introduced to help the European Union meet the Kyoto Protocol targets, and it was hoped that the existence of a market would drive emissions downwards. Although some reductions in overall emissions have been made, it is felt that more could have been achieved if targets had been made more demanding. much of the reduction in emissions that has taken place has been attributed to the economic downturn rather than improvements in efficiency.

As an example of how emissions trading works, consider two companies, A that is able to cut carbon emissions at a cost to itself of £5/tonne CO_2, and B that is able to cut emissions at a cost of £9/tonne.

A sells 1000 tonnes to B at £7/tonne, i.e. £7000, thus making £2000 profit.

B buys 1000 tonnes from A at £7/tonne, saving itself £2000.

Carbon Reduction Commitment (CRC) energy efficiency scheme

The CRC scheme is aimed at cutting those CO_2 emissions in large public and private sector organizations not already involved in CCAs under the CCL and the EUETS. The organizations involved are collectively responsible for about 10 per cent of UK emissions.

For organizations that consume more than 6,000 MWh per year, metered through a half-hourly electricity meter, participation is mandatory, and they must register with the Environment Agency. Annual league tables of performance are produced, and organizations are encouraged to develop better energy management strategies.

Beginning in 2012, participants will buy allowances from the government each year to cover their emissions in the previous year. Thus if they reduce their energy use they can lower their costs. Allowances were initially to be priced at £12 per tonne of CO_2 for the first year and then allowed to vary according to market demand.

CHAPTER 8

CONTROLS AND BUILDING MANAGEMENT SYSTEMS

Heating, ventilation, lighting and air-conditioning systems all require controls so that the spaces which they service enjoy the environmental conditions demanded. Controls are often, and perhaps should be, almost unnoticed by the building user, and automatic control can be a contentious issue. In a modern build-ing it is expected that appropriate conditions will be provided without the need for intervention, but people also like to have personal control over their environment. The optimum solution is often some form of automatic control but with manual over-ride facilities to suit the user's needs. Conventionally, each service had its own control system, but in the last thirty years building management systems (BMS), which can in theory control all the systems under one umbrella, have become stand-ard in most non-domestic buildings.

Before BMS were developed, individual modules were used to control specific functions or items of building services plant; they are still used in smaller buildings and of course in many older buildings. These typically control:

- temperature
- humidity
- time of day switching
- power management
- optimum start/stop
- light switching.

They have certain disadvantages in comparison with BMS. They are not necessarily much cheaper, the data is not available for output elsewhere, and they tend to operate on fixed algorithms with no centralized control and no monitoring and logging facilities.

While a dedicated controller for an air-conditioning system may control the temperature and humidity well, it has certain disadvantages from the point of view of energy efficiency.

- It may not be possible to have a complete overview of all the conditions and settings at any one time.
- There are generally no facilities for recording settings, environmental conditions or operation, to build up a historical record of performance.

These limitations make it difficult to assess the energy efficiency of the system or to tune it for maximum efficiency, but these issues can be addressed by the use of a BMS.

The cost of the energy consumed in non-domestic buildings in the United Kingdom is about £8000 million per annum, rising steeply year by year. BMS have the capability to operate buildings more efficiently to reduce these costs, but often the operators are non-technical personnel with a limited understanding of their potential. A correctly set up and operated BMS can help with automating the operation of a building and improve energy efficiency, but one that is incorrectly used not only costs a great deal to purchase, it could actually waste energy.

BMS are also known as building automation systems (BAS), energy management systems (EMS), energy management and control systems (EMCS), central control and monitoring systems (CCMS), and facilities management systems (FMS). For the sake of simplicity the term BMS is used here to cover all these variations.

The operations that a BMS can perform range from the simplest task such as turning off heating when not required to the performance of complex procedures such as tracking temperatures, adjusting outputs, logging conditions and control sequences. In the hands of a well-trained operator a BMS enables energy usage to be firmly controlled. As well as controlling those functions mentioned above, the BMS might also interface with other control and monitoring services such as safety, access, lifts, security systems and fire security systems, or even be fully integrated with them.

A BMS may control only a few functions in a small building, or many functions in a group of buildings many miles apart. It is important that the BMS is no more complex than required for successful running of the building, as excessively complex systems are a waste of money and tend to be under-used.

Only a few years ago BMS were only considered as an afterthought for many buildings, but if designed in from the beginning of the services design process they should enable greater efficiencies in both energy and operational costs to be achieved.

In the past BMS were only cost-effective in very large buildings, but in recent years the cost in real terms has fallen so much that they are competitive with stand-alone controls in most buildings. Current systems are modular in form, which means that if the building is extended, units can be linked together to control the enlarged building.

In a hard-wired system much of the cost is associated with cabling and installing the sensors, and it is often the case that it is not worthwhile incorporating a BMS into an older building unless a general refurbishment is taking place.

Comfort and environmental control

Heating and lighting systems have long been subject to automatic control, since it is neither convenient nor cost-effective to continually adjust controls on boilers and air-conditioning systems in large buildings. A major stage in automatic control came with the development of central heating provided from a single boiler controlled by a suitably located thermostat. Temperature control of individual rooms became available with thermostatic radiator valves (TRV), which can be manually adjusted to suit individual requirements. The use of gas and liquid fuels, along with electric timers, made automatic control of the boiler easier and provided some flexibility in timing, but it was the advent of physically small and cheap computing systems that heralded the development of BMSs, with fully flexible timetables and multiple zones, and made possible the associated managerial, operational and energy efficiencies.

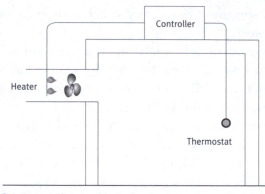

Figure 8.1 Conventional control of a single-zone constant-volume all-fresh air ventilation system using two-position control

Two-position control

In order to understand the benefits of a BMS, it is useful to consider a simple stand-alone control system, an example of which is shown in Figure 8.1. Here a room is provided with ventilation at a constant rate, and a thermostat temperature sensor switches a heater battery on and off in order to maintain a constant temperature. When the thermostat detects a deviation below the set-point temperature, a signal is sent to the actuator in the controller to open a switch that turns the heater battery fully ON (using a two-position ON/OFF control). The information from the signal sent by the thermostat is available only to the actuator, and the switch is controlled only by the information from it. This limited use of information is typical of conventional controls. The only other information controlling the system might be from an overriding timer which limits the ON and OFF periods to certain hours in the day, and switches the fan off and on at those times. A cut-out link prevents the heater from operating when the fan is off.

The simple control system in Figure 8.1 is not particularly good at giving close control of temperature in the space, and is not optimized for energy efficiency. Nor is there the facility for recording the conditions, running hours and energy consumption of the system. The two-position control (ON/OFF) is particularly poor, and is dependent on the electromechanical properties of the components. Solid-state systems are capable of providing much more accurate control of conditions.

The thermostat used to control the heater battery in Figure 8.1 is likely to be of the bimetallic strip type, and switches the heater battery fully ON or OFF. A control differential defines the limits between the 'on' and 'off' points on the thermostat; if it is too small the switching frequency will be too high, leading to excessive wear and accelerated breakdown. An operating differential defines the difference between the highest and lowest temperatures in the room as the heater cycles between on and off, and is larger than the control differential because of the time lags in the system. After the heater is turned off it remains hot for some time, causing the temperature to overshoot on the high side of the set point (see Figure 8.2). The greater the physical distance between the heater and the room, the higher the overshoot will be, because of the increased time lag. In applications such as domestic hot water storage tanks, a large operating differential is not critical, but when heating a room any overshoot represents wasted energy and possible discomfort for the room occupants.

The operating differential can be reduced by employing timed two-position control, where a small heater is built into the thermostat. When the thermostat calls for heat, this heater is energized and the heat generated within the thermostat causes the thermostat to close the heating valve earlier and limit overshooting.

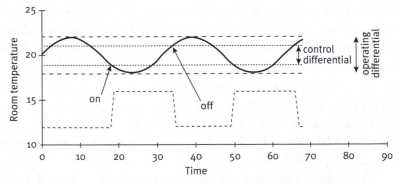
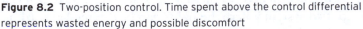

Figure 8.2 Two-position control. Time spent above the control differential represents wasted energy and possible discomfort

Proportional and other controls

In contrast to two-position control, which is 100 per cent ON or 100 per cent OFF, many controls, particularly within BMS, are proportional. That is, a control unit sends a signal to the heater whose magnitude is proportional to the deviation from the set point. As the temperature approaches the set point the output from the heater falls, so overshoots are reduced, and both temperature control and energy efficiency are improved. Other forms of control such as floating, proportional, integral, derived and PID perform better than two-position control, and are found in BMS applications.

Closed and open loops

The system shown in Figure 8.1 is an example of a closed loop system in which the result of the control action (that is, the room temperature) is fed back to the controller. In an open loop system there is no feedback and the result of the control action does not affect the input to the controller. An example of this is heating control based on a thermostat mounted on an outside wall of a building. Its operation is based on the assumption that the heating demand is inversely proportional to the outdoor temperature, which in a general sense is correct, but other factors will influence the heating demand, such as open doors and windows and variations in the number of occupants, and these are not taken into consideration. The outdoor temperature is unaffected by either the state of the heater or the indoor temperature. Clearly such systems used alone have serious limitations and may result in poor control of internal conditions, but they are still found in older buildings. Often, though, outside temperature sensors are used in addition to indoor sensors in a more complex control loop – for frost protection, for example, or in a compensator.

An example of where an open loop system can be useful is a lighting control system where the lights can be dimmed according to the outside light level. The level of daylight outside the building will not be affected by the lighting level inside the building.

System compatibility

In the early days of BMS, components from different manufacturers were not compatible, because of the use of proprietary communications protocols (the 'language' used for communication within the system). The implication of this was that the whole system – sensors, actuators, central control and software – had to be purchased from the same manufacturer. More recently movement towards a truly common communications protocol has been made, and a number of 'open protocols' have been available for some time, including BACnet, Batibus, Lonworks and EIBus. Thus a BACNet compatible valve actuator from one manufacturer will operate correctly with a BACNet compatible sensor from another. KNX is a standard open interface which allows the integration of a number of applications, for which around 200 manufacturers produce compatible devices.

Lighting control is often run from a completely separate system from heating, ventilation and air conditioning (HVAC), and a number of sophisticated lighting controllers use the DALI (digital addressable lighting interface) protocol, which is KNX compatible.

Fire and safety systems may be integrated with the BMS, but it is often felt that a separate fire safety system gives greater protection than one that is fully integrated. If the systems do not use the same protocols they can be integrated through 'gateways' installed for each of the separate services, which can then be programmed to provide a communications network between the separate systems.

Functions of a BMS

The functions provided by a BMS include control of plant, such as:

- automatic switch on/off of heating, ventilation, air conditioning and lighting
- optimization of plant operation and services to minimize energy consumption and improve maintenance

- maximum use of outside air for cooling air-conditioned buildings
- the provision of multiple timetabling and scheduling opportunities, for example weekdays, weekends and holidays.

Functions concerned with monitoring of plant status and system variables include:

- sensing values of important parameters such as temperature, flow rate, humidity, valve position, and plant on/off status
- generating alarms when pre-set values are exceeded
- helping with maintenance by assessing the state of plant (such as the cleanliness of filters)
- taking more rapid remedial action in case of faults, thereby minimizing damage or disruption.

A BMS can be used as an aid to effective maintenance. Routine maintenance increases the potential life of a control system and reduces the frequency of breakdowns and emergency repairs.

The output from the BMS can be linked to a computer-based maintenance system which enables preventive maintenance to be carried out, making it easier for replacement to be made based on the number of hours plant has run, rather than using a fixed period of elapsed time. The switching functions of plant items such as compressors, fans, pumps and boilers can be logged, and the software will total the hours run for each item of plant in order to generate appropriate work orders. This creates an effective preventive maintenance regime under which replacement of parts can be carried out at convenient times with a consequent reduction in breakdowns and unplanned time-outs. It may in fact lead to the amount of maintenance work being reduced, as the servicing is scheduled according to the hours run rather than time elapsed. When an item of plant such as a pump goes beyond its recommended service interval, an alarm report can be generated to remind the staff that it is overdue for servicing. Maintaining items of plant in top condition will enable them to run at their maximum efficiency for longer and save energy in the long run.

A common BMS feature is the use of pressure sensors to monitor the pressure drop across an air filter in the HVAC system. As the filter becomes dirty the pores clog up and the pressure drop across it increases. When it exceeds the set point the BMS will issue an alarm indicating that the filters need changing.

Sometimes an actuator handle shears off the shaft on a three-way valve controlling a heating system; the valve actuator position sensor indicates that it has closed correctly, but the hot water continues to flow through the system and the room temperature rises beyond its set point. Monitoring of the temperatures and sequence of control actions enables the fault to be traced more quickly than without the BMS.

BMS score over stand-alone controls in the provision of energy management information such as monitoring instantaneous energy consumption and temperatures. The extraction of stored historical data and comparison with current values enables trends in consumption to be identified and data for M&T, degree days, CUSUM and other forms of analysis to be generated. Over the longer term, this allows the effectiveness of energy-saving measures to be assessed by 'before' and 'after' interventions.

If a heat meter is fitted the boiler efficiency may also be assessed using BMS data and the method shown in Chapter 5.

Time spent inspecting plant and reading meters manually can be saved, and clocks for an entire installation can be adjusted for summertime change by a few strokes of the keyboard instead of having to visit each plant room separately.

Surveys suggest that only 30 per cent of the capacity of a typical BMS is currently put into use, largely because of insufficient training. Often BMS are set up to produce a large amount of data, much of which goes unanalysed. A survey of 50 managers showed that 82 per cent had BMS but only 2 per cent were able to use them for targeting.

BMS configurations

A BMS requires sensors to measure variables such as temperature and pressure, actuators which will switch the plant on and off and vary the position of valves and dampers, and an intelligent controller. A number of configurations are possible.

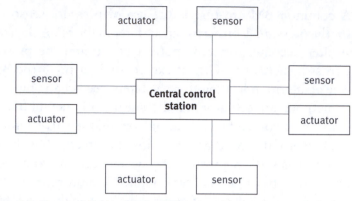

Figure 8.3 Star or radial layout of a BMS

A radial or star layout, typical of older systems, is shown in Figure 8.3. Each sensor and actuator is hard-wired directly to a central control station. This layout is simple and effective, but in larger buildings may require long cable runs, which are expensive and which may result in significant voltage drops, leading to operational difficulties. All the processing will be carried out at the central station.

Modern systems use outstations which gather together local data and control points and permit shorter cable runs (Figure 8.4). Since each outstation possesses a certain level of processing capability, a central station is not necessarily required, as the outstations can be connected together. Access for downloading data or changing setpoints can be made by plugging in a laptop, and can be controlled by the use of passwords. This has major advantages for large organizations such as local authorities and those with sites spread widely over a large area. A central 'building management facility' can control and collect data from all the buildings on the system into one location. It is common now for BMS to be accessible through internet browsers, providing even greater flexibility, and buildings on sites throughout the country can be monitored and controlled from a central facility.

A further BMS configuration is ring topology (see Figure 8.5), in which the network cable is connected to each station in turn and information travels round the ring in one or both directions. This is often used in integrated systems, which may include intruder and access systems, fire alarm and file management systems along with the control of the HVAC equipment.

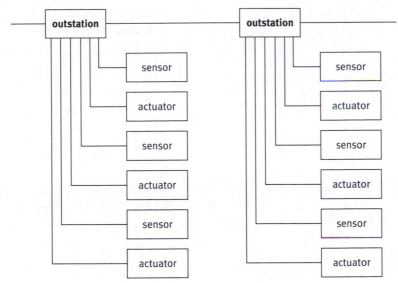

Figure 8.4 Bus connection of multiple intelligent outstations

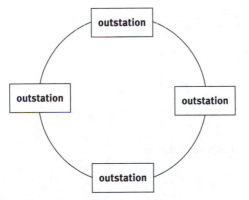

Figure 8.5 Outstations connected in a ring topology

Typical operation of a BMS

A room heater of the type shown in Figure 8.1 may be controlled using a BMS like that shown in Figure 8.6, which shows the main elements of the BMS. These include:

- a central control unit
- outstations
- sensors
- actuators
- cabling.

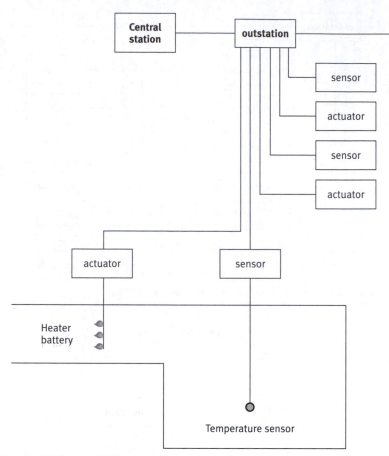

Figure 8.6 Typical BMS components

A signal is sent from a temperature sensor in the room (such as a thermocouple) via the cabling to the outstation. At the outstation, the signal is checked against a previously entered set-point temperature. If the temperature is below the set-point a signal will be sent to the actuator to switch the heater battery to an appropriate level of output. Periodically, signals indicating the room temperature and the heater battery status will be requested by the outstation and central station, and will be transmitted and stored by the central station as required. In distributed intelligence systems the central station will only be involved in setting the set point and in recording the system settings; the major part of the processing is carried out by the outstation.

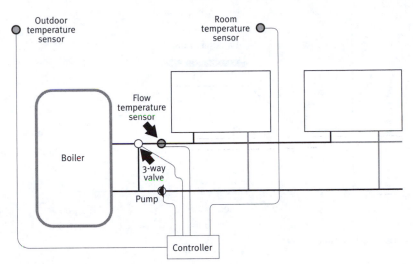

Figure 8.7 Compensated control of a heating system

Compensated on/off control of a heating system

In compensated control the flow temperature of the water flowing through a central heating system is adjusted according to the outside air temperature. If the outside temperature rises, the water flow temperature is reduced, thus allowing lower temperatures to be used in milder conditions (see Figure 8.8).

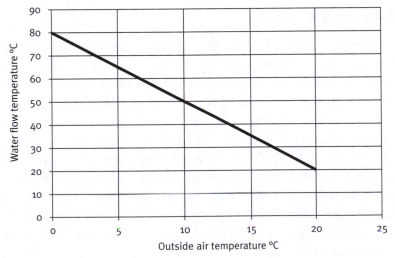

Figure 8.8 Compensator schedule

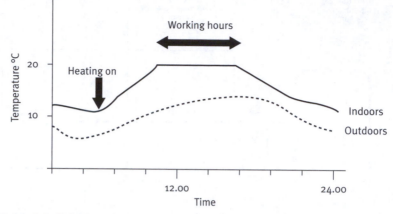

Figure 8.9 Optimum start

The temperature of the boiler flow as sensed by the flow temperature sensor is varied as a function of outdoor air temperature by the controller, which varies the position of the three-way valve and also cycles the burner on and off. The room temperature sensor may also be used to provide room influence control by resetting the compensation ratio. An increase in room temperature would lower the compensation ratio, and a decrease in room temperature would raise the compensation ratio.

Optimum start/stop

Since there is thermal inertia in a heating system (distance–velocity lag) and the building itself, the effects of switching the heating system in a building on or off are not felt immediately. When a heater in a room is switched off it may be half an hour or more before the room becomes uncomfortably cool. In a heavyweight building such as an old stone church there will be a long time lag, but in a lightweight building such as a portable shed it will be very short; it will heat up quickly and cool down quickly. In order for a building to achieve the required temperature at the appropriate time, the heating must be switched on some hours beforehand. How much earlier depends on a number of factors, and in particular:

- the temperature inside
- the temperature outside

- the thermal mass of the building and the type of heating system.

In older buildings without BMS, simple timers were set for the severest conditions and rarely altered subsequently. In milder conditions, which in fact cover most of the heating season, the building reaches the target temperature some hours before opening time, thus wasting energy (see Figure 8.9). There is a maximum rate at which the internal temperature can rise, which will depend on the thermal properties of the building and the heating system, and the outside temperature. Optimum start controllers incorporate an ambient air temperature sensor fixed to the outside wall, an inside temperature sensor, or both. They include algorithms to calculate the appropriate start-up time from a combination of time and temperature inputs, and usually include a frost protection facility.

Optimum stop controllers and optimum start/stop for air conditioning are also available. Building Regulations specify that an optimizer or optimum start program is required for buildings with a space heating load of more than 100 kW.

Electrical load management

Demand side management is so called because customers create a demand for energy which is then supplied mainly by utility companies. To some extent it describes what happens as a result of energy management. Supply side management refers to the various operations carried out by utility suppliers, and is normally designed to enable them to supply power or fuel at the cheapest generation prices, and to set tariffs at a level that recoups the generation prices and provides an appropriate profit. As demand for electricity increases, less efficient generating stations are brought on stream (so the average cost of generation increases) and the price per unit paid by the customer increases accordingly. Per kilowatt hour, electricity is the most expensive fuel in most locations, and accounts for a substantial proportion of the fuel bill in office buildings, which have high air-conditioning and lighting loads. Peak-lopping and load-shifting operations may be programmed through the BMS to reduce high demand penalty charges and avoid peak charging periods.

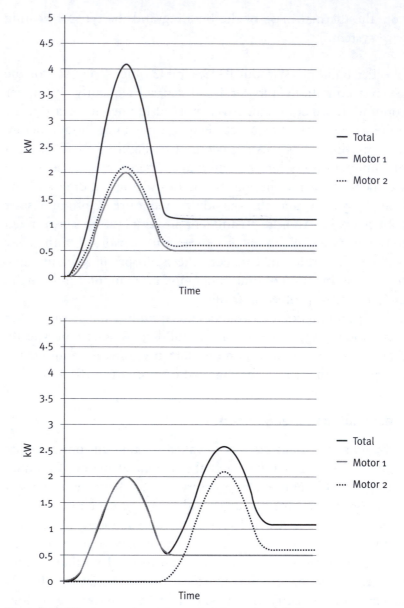

Figure 8.10 Effect on maximum demand of staggering switch-on

Large users of electricity pay a maximum demand charge based on the maximum power used at any one time over a specified period, in addition to the actual electricity units used. For a large organization this can amount to several thousand pounds per month. Maximum demand penalties can be avoided by staggering machine start-ups, and shedding loads according to pre-set priorities: for example, switching off non-essential items of equipment at times of high demand, and so lopping the peak demand and reducing the overcharge. This does not actually save energy but reduces costs to the consumer. In items such as water heaters and refrigeration plant there is considerable thermal inertia, and switching off for short periods results in a non-significant loss of service and may avoid high cost penalties. Demand can be monitored and plant switched off on a priority and size basis.

The example in Figure 8.10 shows how the staggering of the start-up of two electric motors enables the peak current draw to be reduced. The start-up current is high, but demand falls back to a fraction of the peak value after a few seconds. If both motors are started simultaneously peak demand is 4 kW, while if the loads are staggered the peak demand falls to 2.5 kW.

Savings from BMSs

Levermore (2000) surveyed a number of sites to estimate where BMSs were achieving savings. (The results are shown in Table 8.1.) An overall payback period of 5.8 years was achieved. In one building 30 per cent of sensors gave dubious readings because commissioning was rushed.

Table 8.1 System integration

Measure	Percentage saving
Increased heating system efficiency	6.6
Optimum start	5.1
Reduced internal temperature	4.6
Correct holiday settings	4.1
Optimum stop	3.1

The various systems in a building, including BMS, fire, security, and access control, may be integrated in a number of ways, either physically or organizationally. While physical integration has been technically possible for some time, it can create organizational difficulties, such as a clash of functions. For instance, if security is integrated with environmental control, it may not be clear where responsibilities lie, and so is often not implemented. Detailed contractual arrangements need to be put in place from the outset so that the energy manager and others know exactly the extent of their responsibilities.

The advantages of integration are:

- A reduced number of contacts and contracts, for example for maintenance.
- A reduced amount of cabling.
- The integrated system can use a TCP/IP ethernet network for control.
- Reduced cabling/equipment costs, for example through use of shared sensors.
- Reduced staffing levels and therefore costs.
- Increased speed of response and quality of data for decision making.
- Streamlined operation by centralized presentation of information.
- Shared data bases.
- A standardized interface reduces training requirements.
- Improved monitoring of all sensors and similar equipment means that systems should operate better in emergencies, as faults can be more rapidly repaired and overall maintenance standards kept higher.
- Increased flexibility. One sensor may fulfil a number of functions. The integrated system can use sensors from one sub-system for other sub-systems: for example, occupancy sensors used for lighting control can detect the presence of personnel in the building and alert the security sub-system. It may also be desirable to link this in with the fire alarm system, so that if the access system indicates people are working in a particular area, then in the event of a fire, those in charge know to look out for people there.
- Improved energy efficiency. Occupancy sensors or access sensors can also be linked to the HVAC system, so that rooms are only heated or air-conditioned when occupied.

Disadvantages or obstacles to integration are that:

- If one system fails, all systems may fail.
- Traffic loading on the network could cause problems. This is potentially dangerous if the fire alarm system is trying to communicate over an already busy network.
- There could be disputes over job functions and seniority.
- In spec-built offices the precise functions required may not be known.
- There are always 'teething problems' when implementing new technology.
- The interface might become too complicated for efficient operation.
- Downtime for maintenance might affect all systems simultaneously.
- Different protocols in the sub-systems might lead to excessive use of gateways and increase hardware costs.

The BMS can assist with commissioning the building, since it incorporates the sensors and logging capabilities required to check the set-up and operation of the HVAC plant.

CASE STUDY 1
AN OFFICE BUILDING WITH MEDIUM-LEVEL GLAZING

This case study concerns an office building located in an out-of-town business park. It consists of mainly open-plan office space with a small number of meeting rooms. The rooms are spread over two storeys and the net usable area is 3398 m². The heating is controlled by a thermostat in each open-plan space. An air-conditioning system cools the server room and the meeting rooms, amounting to approximately 10 percent of the floor area of the building. Operation of heating and cooling plant is based on the hours of 7.00 to 22.00 each day, the boiler efficiency is estimated at 70 per cent, and a BMS is fitted but only the basic features are utilized.

The energy consumption figures are shown in Table CS1.1, and may be compared with the benchmark figures from ECG 19 in Table CS1.2. A limited energy audit was carried out on the building.

Table CS1.1 A summary of the energy consumption, percentage of total, and costs, for case study 1

Application	Consumption kWh	% of total	Cost £	% of total cost
Heating	671,644	79.8	30,223.98	61.8
Cooling	6,770	0.8	744.7	1.5
Lighting	95,144	11.3	10,465.84	21.4
Other	67,960	8.1	7,475.6	15.3
Total	841,518	100	48,910.12	100.0

Table CS1.2 Good practice and typical annual energy consumption for the building based on benchmarks in ECG 019 (adjusted to allow for A/C for only 10% of the building) Total energy use kWh /year

Application	Good Practice	Typical
Heating + hot water	329,606	604,844
Cooling	4,757	10,533
lighting	91,746	183,492
Other electricity	101,940	132,522
Total	528,049	931,391

The figures show that the performance is better than typical, but falls short of best practice. Thus, there is room for some improvements to be made. Closer inspection of the figures shows that the potential areas of improvement are heating and hot water, and lighting.

The U-values of the elements, from the original building data, are as shown in Table CS1.3. The glazing ratio (GR) is 50 per cent.

The lighting level throughout the offices is 300 lux, which is an appropriate level for open-plan offices but is not particularly efficient at 20 W/m^2. This gives a total installed wattage of 69,960 W, which, allowing for a diversity factor of 0.5 gives an annual energy consumption for the lighting of 95,144 kWh. Other electrical use is estimated at 20 kWh/m^2/year, giving an annual total of 67,960 kWh (11.4 per cent of total energy use).

The building is located in Scotland where heating degree days = 2500 and cooling degree days = 108.

Applying correction factors of 0.75 for five-day week use and 0.6 for intermittent plant operation, the annual heating energy consumption = $17413 \times 0.75 \times 0.6 \times 0.024 \times 2500/0.7 = 671,644$ kWh/year (80 per cent of total energy use).

Allowing for a coefficient of performance (COP) of 3.0 for air conditioning, the cooling energy is $17413 \times 0.75 \times 0.6 \times 0.024 \times 108/3 = 6770$ kWh/year (0.8 per cent of total energy use).

A summary of the energy consumption, percentage of total, and costs, is given in Table CS1.1.

Table CS1.3 Case study 1 energy consumption figures

Element	U-value (W/sq.m.K)	Area (sq.m)	UA (W/K)	% Fabric heat load	% Total heat load
Walls	0.84	480	403.2	5.5	2.3
Glazing	5.6	480	2,688.0	36.7	15.4
Roof	0.49	1,699	832.5	11.4	4.8
Floor	2	1,699	3,398.0	46.4	19.5
		ΣUA	7,321.7		
Ventilation conductance		(0.33nV)	10,092.0		58.0
		TLC	17,413.7		

The lighting constitutes over 20 per cent of the total energy costs; improvements could involve improved control and/or the use of more efficient tubes and luminaires. Short-term low-cost measures to improve the lighting could include stickers next to light switches. Longer-term measures requiring significant investment involve changing the lighting control strategy to include occupancy sensing or daylight level sensing, but it would be more appropriate to instigate these during a comprehensive lighting replacement programme. The long-term strategy would include an investigation into the costs of such a programme.

The heat loss from the curtain walling represents only 5 per cent of the fabric heat load and 2.3 per cent of the total heating demand, therefore any improvements are unlikely to prove cost-effective. A reduction of the U-value by 50 per cent, resulting in a lowering of the energy consumption by only 2.5 per cent, would save only £130 per year at a cost of over £5000, a payback period of over 40 years. A change from single to double glazing saves £750 per year but at a payback period of over 20 years. This could be justified if there were other reasons for changing the glazing, such as rotting or corroded frames, or for purposes of soundproofing. At this particular out-of-town location sound pollution is not a problem and the latter does not apply. The double glazing would have the further effect of reducing draughts and infiltration, which constitutes a further heat loss. Taking this into account would reduce the payback period by another five to six years, but it remains an unattractive proposition.

Specific recommendations

- If double glazing is not to be implemented, then some form of draught stripping would reduce infiltration, with a payback of three to four years.
- Addition of thermostatic radiator valves (TRV) would improve control and heating efficiency at a cost of £10–15 per radiator.
- A rolling programme of installing sub-meters could be paid for from the energy savings.
- Investigate the possibility of providing a separating chiller for the server room so that cooling operation can be optimized.
- Install blinds to reduce building cooling load.
- Organize an information campaign to urge staff to turn off equipment such as computers, photocopies and printers at night.
- The use of the BMS should be extended to include:
  - optimum sequencing of boilers
  - night set-back for heating
  - optimum start/stop for heating
  - maximum use of outside air for cooling.
  - Limit basic hours of operation to 9.00–18.00 Monday to Friday, and instigate a booking system for provision of services outside these hours.
  - Set the BMS to log performance and use the results for monitoring and targeting.
  - This would require investment in training the building manager in the use of the BMS, but is likely to give very quick returns.
- The long-term strategy should include:
  - Establish a database of energy consumption figures and related records.
  - Identify the company's strategic plans and investment criteria.
  - Establish an energy efficiency programme.
  - Establish a replacement programme for lighting.

The specific recommendations for improvement, particularly for heating and cooling, are highly dependent on location. If the same building were located in London, where heating degree days and cooling degree days are 2129 and 365 respectively, the heating load becomes 75 per cent (cost, 55 per cent) and cooling load 3 per cent (cost, 5.4 per cent) of the total. In Rome, with heating and cooling degree days of 1103 and 1173 respectively, the heating load falls further to 55 per cent (cost, 33 per cent) while the cooling load increases to 13.8 per cent (cost 20 per cent).

Table CS1.4 Summary of energy consumption and costs for different locations

Application	Consumption kWh	% of total	Cost £	% of total cost	Edinburgh
Heating	671,644	79.8	30,223.98	61.8	
Cooling	6,770	0.8	744.7	1.5	
Lighting	95,144	11.3	10,465.84	21.4	
Other	67,960	8.1	7,475.6	15.3	
Total	841,518		48,910.12		
Application	**Consumption kWh**	**% of total**	**Cost £**	**% of total cost**	**London**
Heating	571,972	75.5	25,738.74	55.7	
Cooling	22,873	3.0	2,516.03	5.4	
Lighting	95,144	12.6	10,465.84	22.7	
Other	67,960	9.0	7,475.6	16.2	
Total	757,949		46,196.21		
Application	**Consumption kWh**	**% of total**	**Cost £**	**% of total cost**	**Rome**
Heating	296,234	55.6	13,330.53	33.9	
Cooling	73,508	13.8	8,085.88	20.5	
Lighting	95,144	17.9	10,465.84	26.6	
Other	67,960	12.8	7,475.6	19.0	
Total	532,846		39,357.85		

CONVERSION OF A TRADITIONALLY BUILT DWELLING TO OFFICE USE

This case study concerns the adaptation of an old dwelling into a small office premises for a software development company. The property is a large house about 100–150 years old, situated in a rural location in east-central Scotland. It is a traditional stone-built house with a slate roof, internal plastering, but no insulation, and a suspended timber ground floor. There are single-glazed hardwood-framed sash windows. It is heated by a very old and inefficient central heating system with large cast-iron radiators, the boiler being highly corroded and unsafe. The house was occupied by an elderly couple and has not been modernized; the electrical wiring has not been replaced for over 40 years. The water supply and drainage pipes are intact, but new lagging is required on the cold water supply. The house has been empty for two years, during which time there was a very bad winter and some of the central heating pipes burst, as did the hot water supply pipe to the taps. It has been purchased by a small financial services company that wishes to convert it to office premises. There is some land in front of the house that could be used for car parking and a yard to the rear of the premises.

The usable floor area is 320 m², spread over three floors. Initially the house comprised sixteen occupied rooms, including hallways, but it will be possible to remove some internal partitions to create larger office spaces.

The accommodation required comprises offices for ten staff, each with a desktop computer, and with one photocopier and two printers in total, along with a small rest room with a kettle and microwave oven. Staff toilets will be required. The working hours will be 9.00–17.30 Monday to Friday.

The first task is to establish the performance of the building as it stands. Unfortunately utility bills are not available, and the performance is to be estimated using the known characteristics of the building.

Using values of thermal properties from the CIBSE guide Part A, the U-values shown in Table CS2.1 were estimated, and the areas assessed from a survey of the building.

Table CS2.1 U-values for the original building fabric, case study 2

	U (W/ sq.m.K)	A (sq.m)	U.A (W/K)	% Fabric loss	% Total heat loss
Roof	2.3	190	437	25.1	19.9
Floor	0.9	170	153	8.8	7.0
Walls	2.3	400	920	52.9	41.8
Windows	5.7	40	228	13.1	10.4
		ΣUA	1,738		
Infiltration = 2.0 ac/hr					
Ventilation conductance =			462	W/K	
		TLC	2,200	W/K	

The sash windows are ill-fitting and allow draughts to enter, and the original flues allow some ventilation. The infiltration rate has therefore been set at two air changes per hour.

Ventilation conductance is therefore $0.22 \times 2 \times 700 = 462$ W/K.

The total conductance is $1738 + 462 = 2200$ W/K.

Using a twenty-year average degree day value for the location of 2500 DD, the annual space heating demand is 132,000 kWh.

Allowing a factor of 0.7 for intermittent heating, and 60 per cent for the boiler efficiency, this equates to an energy requirement of 154,000 kWh, or 481.25 kWh/m². This compares extremely unfavourably with the 'good' and 'typical' values of 79 and 151 quoted in the CIBSE Guide part F. At approximately six times the 'good' value, there is clearly huge scope for improving the heating performance of the building.

Hot water and cooking energy consumption are each estimated at 3000 kWh per year.

The other main supply of energy is electricity. In its original use as a dwelling, electricity was used mainly for lighting and in the central heating pump. A quick assessment is made below.

Lighting three rooms at a time each with a 100 W tungsten bulb, for four hours a day throughout the year, gives annual consumption of 432 kWh/year. Energy needed for the central heating pump, assuming a 150-day heating season, and the use of 2kW for six hours a day, is 1800 kWh (see Table CS2.2).

Table CS2.2 Estimated original energy consumption and costs

Application	kWh	% Energy	Cost £	% cost
Space Heating	154,000	94.92579	6,930	93.08261
Water heating	3,000	1.849204	135	1.813298
Cooking	3,000	1.849204	135	1.813298
Lighting	432	0.266285	47.52	0.638281
Other	1,800	1.109522	198	2.659503
Total	162,232		7,445.52	

The total electricity consumption amounts to 7 kWh/m^2 per year, far below the figures for a 'good' office. However a direct comparison is not realistic, as far more electricity will be consumed when it is in use as an office. There will be a desktop computer for each of the ten employees, and a photocopier and two printers. The house should be rewired as a matter of course, and the wiring upgraded to allow for greater consumption. At the same time data cabling can be installed for a local area network (LAN) and internet connections. A server will be required and can be housed in one of the refurbished outhouses; with appropriate ventilation, cooling will not be required for the server. Compact fluorescent lighting will produce a high level of savings. Table CS2.3 gives an estimate of the new electricity consumption.

Table CS2.3 Electricity consumption in the converted building

Item	Mean consumption W	Hrs/year	kWh/year
Computer × 10	2,000	2,000	4,000
Lighting 10W/sq.m	3,200	1,000	3,200
Kettle	3,000	220	660
Microwave	1,000	240	240
Photocopier	300	250	75
Printer × 2	400	250	60
		Total	8,235

This gives a value of 16.7 kWh/m², significantly better than the 'good' value.

Options for improving the building

Since the heating is clearly the largest consumer of energy, most of the emphasis will be placed on improving the insulation of the building and supplying the heat from more efficient plant. Consideration should be given not only to economics but also to the practical aspects of adding insulation – in particular its thickness, since the insulation will be on the inside of the walls.

Table CS2.4 suggests reasonable options for the building fabric.

Table CS2.4 Proposals for the building fabric

Measure	Thickness mm	U value W/m²K
Roof insulation below rafters	140	0.25
Floor - board beneath joist	200	0.16
Walls - rigid board + plaster	100	0.35
Double glazing		2.2

This produces the figures shown in Table CS2.5.

Table CS2.5 Energy requirements of the altered building

	U W/m²K	A m²	UA W/K	
Roof	0.25	190	47.5	
Floor	0.16	170	27.2	
Walls	0.35	400	140	
Windows	2.2	40	88	
		ΣUA	302.7	
Infiltration = 1.0 ac/hr				
Ventilation conductance =			231	W/K
		TLC	533.7	
AHL	32,022			
Using condensing boiler at 88% efficiency				
AED	25,472.05	kWh		
per m²	79.60014	kWh/m²		

As there is no mains gas, the cost of gas will be slightly higher and the cost of the storage tank must be factored in, but it remains the cheapest fuel option for this location. The heating performance is in the 'good' range mentioned earlier.

The natural choice of heating system would be either a traditional radiator system or underfloor heating. Underfloor heating requires a greater upheaval to the building, but as the building is empty and is to be generally refurbished this will not increase the costs significantly and can be fitted easily into the work schedule. It will take slightly longer to heat up an a cold Monday morning after the office has been closed for the weekend, but can be used effectively with high-efficiency sources of heat such as a condensing boiler or heat pump. Incorporation of optimum start/stop controls will enable maximum thermal comfort and efficiency to be achieved.

Thermostatic radiator valves should be used for local control of heating.

In such a small building a building management system (BMS) is not worthwhile. A good programmable controller should be installed for the heating. Time switches or motion detectors could be used effectively with the lighting, since the wiring and controls will be renewed in any case. For domestic hot water, point-of-use heaters should be effective, since the quantity of hot water required is limited, and this obviates the necessity for renewing the hot water pipework to the taps.

Table CS2.6 shows the new estimated energy consumption for the converted building, and the distribution of energy costs. The heating costs have fallen dramatically, but the electrical use has increased due to the use of computers and other office equipment.

Table CS2.6 New estimated energy consumption and distribution of energy costs

Application	kWh	% Energy	Cost £	% cost
Space heating	25,472	71.33615	1,146.24	53.51261
Water heating	2,000	5.601143	90	4.201681
Cooking	0	0	0	0
Lighting	3,200	8.961828	352	16.43324
Other	5,035	14.10088	553.85	25.85668
Total	35,707		2,142.09	

Table CS2.7 gives approximate costs and payback periods for the options recommended above.

Table CS2.7 Approximate costs and payback periods

Element	Cost £	Annual saving £	Payback years
Walls	4,779	2,387	2.0
Roof	6,399	1,176	5.4
Floor	4,266	568	7.5
Windows	14,000	428	32.7

The boiler has not been costed as the existing one is in any case unusable; the same applies to the lighting installation.

Other options

These include the following.

Ground or air source heat pump

In the absence of mains gas, heat pumps present an efficient alternative for space heating. For a ground source heat pump

(GSHP), space would be available beneath the car park or back yard for burying the coils.

Capital cost

- GSHP approx £7,000 including excavation for coils. Mean COP 4.0.
- Air source heat pump (ASHP) approx £5,000. Mean COP 3.0.
- Electrical input for GSHP = 22,415/4 = 5603 kWh. Annual heating cost £616.30
- Electrical input for ASHP = 22,415/3 = 7471 kWh. Annual heating cost £821.81

Compare this with £1146 for a gas condensing boiler. The initial cost of the GSHP will be higher, since excavation is needed for the underground heating coils, probably not less than £7000. Payback compared with condensing boiler = 6000/530 = 11.3 years.

Payback on ASHP = 5000/324 = 15 years.

Although these payback periods are rather long, the maintenance required on heat pumps is lower. The optimum choice of heating system would be underfloor heating, which would allow low water temperatures and hence maximum COP to be obtained.

Renewables

Solar hot water is also a possibility. As an office, the hot water consumption will be limited to hand-washing, and will amount to no more than 6000 kWh per year. A panel 3 m × 3 m (9 m²) would produce about half of that amount, and a supplementary heating source would still be required to top up, and for dull days. When the cost of plumbing and connecting to the supplementary source are included, it is unlikely that a system could be sourced for less than £5000, which would give a payback of about twenty years. The renewable heat initiative (RHI) would reduce this by only about two years.

Solar photovoltaics (PV)

Thanks to the feed-in tariffs (FIT) solar PV has become more attractive economically. A 1 kWp panel would require about 8 m^2 and would produce about 800 kWh per year.

The maximum panel size for the roof would be approximately 50 m^2, producing 7 kWp. This would produce a net saving, when taking into account income from exported electricity, of £2100, and because of the FIT the payback is reduced to only thirteen years.

A biomass boiler would also be an option for such a rural location, and a detailed study would include sourcing and pricing local biomass fuel. Storage space would also be required (and is available at the back of the house).

ESTIMATING ENERGY CONSUMPTION USING DEGREE DAYS

As heating is the main consumer of energy in buildings in the United Kingdom, it is important to acquire at least an approximate estimate of the amount of energy it consumes. In this section the concept of degree days is introduced, and a simple method of using them to make a rough estimate of annual heating energy consumption is presented.

If building plans and specifications are available, then the overall thermal performance of the building can be calculated, and the annual heating energy consumption estimated using computer-based simulation models or by using the calculation methods shown below. It is a fairly simple matter to set up a spreadsheet which will make repeated calculations easier, so that the effect of making a range of changes to the building can readily be assessed.

Consider a building maintained at a temperature of 20°C over a 24-hour period. The rate of energy input required to maintain that temperature depends on the indoor–outdoor temperature difference, and the thermal properties of the building, characterized by the sum of the fabric and ventilation conductances, here defined as the total loss conductance (TLC).

The fabric conductance accounts for conduction heat losses through the walls, windows, doors, roof and ground floor of the building, and is calculated by multiplying the U-value of each element of the outer envelope of the building by its area.

Example

Walls of U-value 0.45 W/m²K, total area 120 m², UA (U-value of building multiplied by area) = 0.45 × 120 = 54 W/K

UA	(W/K)
Walls	54
Roof	37.5
Windows	80
Floor	30
ΣUA	201.5

The ventilation conductance is based on the flow rate m in kg/s of cool air drawn into the building from outside, and is equal to mC where C is the specific heat in J/kgK. Inserting appropriate values of density and specific heat, this reduces to 0.33 nV where n is the ventilation rate in air changes per hour and V is the internal volume of the building in m³.

Thus, for V = 225 and n = 2, 0.33nV = 0.33 × 2 × 225 = 148.5 W/K.

$$TLC = \Sigma UA + 0.33nV$$
$$TLC = 201.5 + 148.5 = 350.0 W/K$$

At any given moment the heat input required is equal to TLC × indoor–outdoor temperature difference (ΔT). It should be noted that where the air and radiant temperatures inside the building are substantially different, this simplification introduces an error in excess of 3 per cent. For many buildings the uncertainty in the input data will be greater than this.

Thus the average heating demand (in W) over a day is equal to $TLC \times \Delta T_{ave}$ where ΔT_{ave} is the average temperature difference over the day.

The overall heating demand in kWh is given by

$$\Delta T \cdot TLC \cdot 24/1000 \text{ kWh}.$$

Assume that the required indoor temperature T_{in} is 20 °C and that outdoors T_{out} is 7 °C:

$\Delta T = 20 - 7 = 13K$

Total heating demand for the day = $13 \times 350 \times 0.024$ = 109.2 kWh.

The above calculations assume there are no other sources of heat within the building apart from the heating system, but this is not the case – incidental gains, such as those from the occupants (about 100 W of sensible heat for a person sitting at a desk working), solar radiation, and electrical equipment such as lights and computers all supply heat which will raise the indoor temperature 'free', and will therefore reduce the amount of heating that needs to be put in through the heating system. The indoor temperature required is reduced by an amount equal to that provided by the "free" gains increase to produce a *base temperature* T_b; the standard value used in the UK is 15.5 °C, implying that 4.5 °C of heat are provided from sources other than the heating system. The degree-day values to various base temperatures are shown in Table A1. The difference between the base temperature and the mean outdoor temperature for any given day is defined as the number of degree days for that day. Where T_{out} is higher than T_b, zero degree days are recorded.

Thus, using degree days in the above example, the effective ΔT is given by:

$$\Delta T = T_b - T_{out} = 15.5 - 7 = 8.5K.$$

And the actual heating demand for the day = $8.5 \times 350 \times 0.024$ = 71.4 kWh.

To estimate the annual heating energy consumption it would only be necessary to sum the individual daily heating demand values; however, in the degree-day method the degree days for each day are summed to give an annual degree-day value (Dd), whose units are degrees × days. The degree days in effect represent the cumulative temperature difference over the year.

For the example above at a location in East Anglia (2254 DD from Table A2):

Annual heating demand = $2254 \times 350 \times 0.024$ = 18,933 kWh

The value of T_{out} used is the mean dry-bulb temperature for that day, but if this is unavailable, the average of the daily maximum and minimum temperatures may be used. Weekly and monthly degree-day totals are useful for detailed energy analysis, as shown in Chapter 3.

In practice, it is not always necessary to calculate the degree days from weather data at individual locations, as they are tabulated in the CIBSE *Guide A*, in journals and through various internet sites for a range of base temperatures.

This is a simplification, and assumes constant heating over the heating season, ignores the differences between radiant and warm air heating, and also assumes a particular level of internal and solar gains (the basis of T_b).

Correction factors can be applied to allow for intermittent use. The actual amount of fuel used can be estimated by taking into account the efficiency of the boiler; the annual heating demand is divided by the fractional efficiency.

In the example used, a boiler efficiency of 75 per cent would result in an annual energy use of 20,588/0.75 = 27,451 kWh.

With gas having a calorific value of 38.7 MJ/m³ this amounts to an annual gas consumption of 27,451 × 3.6/38.7 = 2553 m³ (1 kWh = 3.6 MJ).

Cooling degree days

It is also possible to calculate cooling degree days which enable the annual cooling load to be determined, but different base temperatures may be used as shown in Table A1. Using T_b =15.5 °C as before, typical annual cooling degree days for London and Edinburgh would be 365 and 108 respectively. The example in Case study 2 shows how cooling loads are calculated.

Table A1 Monthly heating degree-day and cooling degree-hour totals to various base temperatures: London (Heathrow) (1982–2002)

Base temp °C	Monthly heating degree-days (K·day) for stated base temperature											
	Jan	Feb	Mar	Apr	May	Jun	Jul	Aug	Sep	Oct	Nov	Dec
10	150	140	99	61	16	2	0	0	4	22	84	132
12	207	192	151	101	37	8	1	2	11	46	130	187
14	267	247	208	150	72	24	6	8	28	86	184	246
15.5	314	290	255	192	105	45	16	18	51	124	228	293
16	329	304	269	206	117	52	20	23	59	135	243	307
18	391	360	331	264	168	91	45	50	100	192	302	369
18.5	406	373	345	277	182	102	55	58	113	207	317	384
20	453	417	393	323	224	138	82	87	152	253	362	431

Base temp °C	Monthly cooling degree hours for stated base temperature											
	Jan	Feb	Mar	Apr	May	Jun	Jul	Aug	Sep	Oct	Nov	Dec
5	1,347	1,216	2,166	3,236	5,935	7,820	9,965	9,630	7,232	5,101	2,507	1,622
12	8	20	109	443	1,626	2,972	4,787	4,467	2,454	962	158	43
18	0	0	2	32	274	635	1,388	1,158	308	33	0	0

Source: CIBSE Guide A. Table 2.23. Reproduced by kind permission of the Chartered Institution of Building Services Engineers.

Table A2 Mean monthly annual heating degree-day totals (base temperature 15.5°C) for 18 UK degree-day regions (1976–95)

	Degree-day region	Mean total degree-days (K·day)												
		Jan	Feb	Mar	Apr	May	Jun	Jul	Aug	Sep	Oct	Nov	Dec	Year
1	Thames Valley (Heathrow)	340	309	261	197	111	49	20	23	53	128	234	308	2,033
2	South-eastern (Gatwick)	351	327	283	218	135	68	32	38	75	158	254	324	2,255
3	Southern (Hurn)	338	312	279	222	135	70	37	42	77	157	246	311	2,224
4	South-western (Plymouth)	286	270	249	198	120	58	23	26	52	123	200	253	1,858
5	Severn Valley (Filton)	312	286	253	189	110	46	17	20	48	129	217	285	1,835
6	Midland (Elmdon)	365	338	291	232	153	77	39	45	85	186	271	344	2,425
7	W. Pennines (Ringway)	360	328	292	220	136	73	34	42	81	170	259	331	2,228
8	North-western (Carlisle)	370	329	309	237	159	89	45	54	101	182	271	342	2,388
9	Borders (Boulmer)	364	328	312	259	197	112	58	60	102	186	270	335	2,483
10	North-eastern (Leeming)	379	339	304	235	159	83	40	46	87	182	272	345	2,370
11	E. Pennines (Finningley)	371	339	294	228	150	79	39	45	82	174	266	342	2,307
12	E. Anglia (Honington)	371	338	294	228	143	74	35	37	70	158	264	342	2,254
13	W. Scotland (Abbotsinch)	380	336	317	240	159	93	54	64	107	206	286	358	2,494
14	E. Scotland (Leuchars)	390	339	320	253	185	104	57	65	113	204	290	362	2,577
15	NE Scotland (Dyce)	394	345	331	264	194	116	62	72	122	216	295	365	2,668
16	Wales (Aberporth)	328	310	289	231	156	89	44	44	77	156	234	294	2,161
17	N Ireland (Aldergrove)	362	321	304	234	158	88	56	56	102	189	269	330	2,360
18	NW Scotland (Stornoway)	336	296	332	260	207	124	88	88	135	214	254	330	2,671

Source: CIBSE Guide A. Table 2.17. Reproduced by kind permission of the Chartered Institution of Building Services Engineers.

ADDITIONAL DATA AND CALCULATIONS

Bills from the utility company will state the quantity of fuel used, which may be in units of kWh, cubic metres or litres, for example. It is convenient to convert everything to a common unit, such as kilowatt hours (kWh). Some conversion factors are given in Table A3.

Table A3 Conversion factors for energy units

Fuel	Gross Calorific value (CV)	CV in kWh	Quantity /KWh
Natural gas	38.7 MJ/m³	10.75 kWh/m³	0.09 m³
Medium oil	40.9 MJ/litre	11.36 kWh/litre	0.088 litres
Propane	93 MJ/m³	25.8 kWh/m³	0.038 m³
Butane	122 MJ/m³	33.8 kWh/m³	0.029 m³
Coal	27.4 MJ/kg	7.6 kWh/kg	0.131 kg

Note: 1 kWh = 3.6 MJ

Carbon dioxide emission factors on the basis of gross calorific value (CV) are shown in Table A4.

Table A4 Carbon dioxide emission factors on the basis of gross calorific value

Energy source	kg CO_2/kWh
Grid electricity	0.54522
Natural gas	0.18523
LPG	0.21445
Coal	0.32227
Wood pellets	0.03895
Diesel	0.25301
Petrol	0.24176
Fuel oil	0.26592
Burning oil	0.24683

Source: courtesy of the Carbon Trust.

Benchmark allowances for internal gains in typical buildings

Table A5 Benchmark allowances for internal heat gains in typical buildings

Building type	Use	Density of occupation m²/person	Sensible heat gain Wm⁻²				Latent heat gain Wm⁻²	
			People	Lighting	Equip		People	Other
Offices	General	12	6.7	8-12	15		5	–
	City centre	16	5	8-12	12		4	–
		6	13.5	8-12	25		10	–
		10	8	8-12	18		6	–
	Trading/dealing	5	16	12-15	40		12	–
	Call centre/floor	5	16	8-12	60		12	–
	Meeting/conference	3	27	10-20	5		20	–
	IT rack rooms	0	0	8-12	200		0	–
Airports/stations	Airport concourse	0.83	75	12	5		50	–
	Check-in	0.83	75	12	5		50	–
	Gate lounge	0.83	75	15	5		50	–
	Customs/immigration	0.83	75	12	5		50	–
	Circulation spaces	10	9	12	5		6	–
Retail	Shopping malls	2-5	16-40	6	0		12-30	–
	Retail stores	5	16	6	0		12	–
	Food court	3	27	10	*		20	?

Cont'd

Cont'd from p156

Building type	Use	Density of occupation m²/person	Sensible heat gain Wm⁻²				Latent heat gain Wm⁻²	
			People	Lighting	Equip		People	Other
	Supermarkets	5	16	12	*		12	?
	Department stores							
	Jewellery	10	8	55	5		6	–
	Fashion	10	8	25	5		6	–
	Lighting	10	8	200	5		6	–
	China/glass	10	8	32	5		6	–
	Perfumery	10	8	45	5		6	–
	Other	10	8	22	5		6	–
Education	Lecture theatres	1.2	67	12	2		50	–
	Teaching spaces	1.5	53	12	10		40	–
	Seminar rooms	3	27	12	5		20	–
Hospitals	Wards	14	5.7	9	3		4.3	–
	Treatment rooms	10	8	15	3		6	–
	Operating theatres	5	16	25	60		12	–
	Hotel reception	4	20	10-20	5		15	–
	Banquet/conference	1.2	67	10-20	3		50	–
	Restaurant/dining	3	27	10-20	5		20	–
	Bars/lounges	3	27	10-20	5		20	–

Source: CIBSE Guide A Table 6.2. Reproduced by kind permission of the Chartered Institution of Building Services Engineers.

Discount factor calculation

See Chapter 6. The discount factor is calculated according to the following formula.

$$F_{(T)} = \frac{1}{(1 + r)^T}$$

Where

$F_{(T)}$ = the factor by which future cash flow at time T years from now must be multiplied to obtain the present value.
r = fractional discount rate.
T = Time from the present in years.

Operative and dry resultant temperatures

A widely used temperature measurement for assessing the thermal comfort of humans was the dry resultant temperature, defined as

$$t_{res} = 0.5_{tr} + 0.5\ t_{ai}$$

where

t_{res} is the dry resultant temperature
t_r is the mean radiant temperature
t_{ai} is the dry bulb air temperature
and where the air speed is less than 0.1 m/s.

A concept more widely used now is the operative temperature, which at the low air speeds normally encountered inside buildings is equal to the dry resultant temperature.

SELECTED BIBLIOGRAPHY

Awbi, H. B. *Ventilation in Buildings*, 2nd edition, Spon, 2003.

Beggs, C. *Energy: Management, Supply and Conservation.* Butterworth-Heinemann, 2002.

British Standards Institution (BSI). *Energy Management Systems – Requirements with guidance for use.* BS EN 16001:2009.

Building Research Energy Conservation Support Unit (BRECSU). Energy Consumption Guide ECG 019, 'Energy Use in Offices.' BRECSU. 2003.

Chang, Wei. *Intelligent Buildings and Building Automation.* Spon, 2010.

Chartered Institution of Building Services Engineers (CIBSE). *Guide A: Environmental Design.* 2006.

CIBSE. *Guide B: Heating, Ventilation, Air Conditioning and Refrigeration.* 2002.

CIBSE. *Guide F: Energy Efficiency in Buildings.* 2004.

CIBSE. *Guide H: Building Control Systems.* 2009.

CIBSE. *Natural Ventilation in Non-Domestic Buildings, Applications Manual AM10.* 2005.

CIBSE. *Mixed-mode Ventilation, Applications Manual AM13.* 2000.

CIBSE. *Managing Your Building Services, Knowledge Series KS02.* 2005.

CIBSE. *Refurbishment for Improved Energy Efficiency, Knowledge Series KS12.* 2007.

CIBSE. *Energy Efficient Heating, Knowledge Series KS14.* 2009.

CIBSE. *Testing Buildings for Air Leakage, Technical Manual TM23.* 2000.

CIBSE. *Building Energy Metering, Technical Manual TM39.* 2006.

CIBSE. *Degree Days, Technical Manual TM41.* 2006.

CIBSE. Energy Benchmarks, Technical Manual TM46. 2008.

Gevorkian, P. *Sustainable Energy Systems Engineering*. McGraw-Hill, 2006.

Levermore, G. J. *Building Energy Management Systems*, 2nd edition. Spon, 2000.

Moss, K. J. *Energy Management in Buildings*. E & F Spon, 2006.

Underwood, C. P. *HVAC Control Systems: Modelling, Analysis and Design*. Spon, 1999.

Journals

Energy World. Available to members of the Energy Institute.
Energy in Buildings and Industry: www.eibi.co.uk
CIBSE Journal. Available to members of the CIBSE.
Energy & Environmental Management:
www.eaem.co.uk

Useful organizations

Building Research Establishment (BRE), Garston, Watford, Herts WD2 7JR. 01923 664258.
www.bre.co.uk

Building Services Research and Information Association (BSRIA). Old Bracknell Lane West, Bracknell, Berks RG12 7AH.
www.bsria.co.uk

Chartered Institution of Building Services Engineers (CIBSE). Delta House, 222 Balham High Road, London SW12 9BS. 0181 675 5211.
www.cibse.org

The Carbon Trust administers many of the British government's energy efficiency programmes and provides advice and literature, much of it free, aimed at commercial buildings and the public sector.
www.carbontrust.co.uk

The Energy Saving Trust also provides energy saving advice and is aimed principally at the domestic sector.
www.energysavingtrust.co.uk.

The Energy Institute (EI). 61 New Cavendish Street, London W1G 7AR. 0171 580 7124. The professional body for those working in energy.
www.energyinst.org.uk

INDEX